Simple Chemistry Investigations

Solutions for doing science in the classroom.

Christopher P. Garside

Seven Sides Publishing

Seven Sides Publishing has a mission to improve teaching and the understanding of science. To contact us, send an email to simpleinvestigations@sevensidespublishing.com or visit our website at sevensidespublishing.com.

Copyright © 2021 by Seven Sides Publishing and Christopher P. Garside. All rights reserved. No part of this publication may be reproduced, stored in a retrieval system, scanned, or transmitted in any form or by any means, electronic, mechanical, photocopying, recording, or otherwise, without the prior written permission of Seven Sides Publishing and Christopher P. Garside. Photocopying is permitted from this book to make copies only for the students of the teacher who owns this book; this is not for other teachers, students, or anyone else in the school or district's students.

ISBN: 9798671484670

Published by: Seven Sides Publishing, Cypress, TX.

Table of Contents

Introduction	Page 4
Unit 1 Properties of and Changes in Matter	Page 17
Unit 2 Gas Laws	Page 30
Unit 3 Energy, Phases of Matter, and Calorimetry	Page 40
Unit 4 Pure Substances and Mixtures	Page 57
Unit 5 Atomic Structure and Nuclear	Page 80
Unit 6 Electrons	Page 109
Unit 7 Periodic Table	Page 127
Unit 8 Chemical Bonds	Page 140
Unit 9 Energy in Chemical Changes	Page 165
Unit 10 Classification of Chemical Reactions	Page 174
Unit 11 Stoichiometry	Page 192
Unit 12 Acids and Bases	Page 206
Chemistry and IPC TEKS & NGSS Correlations	Page 220
Equipment List for all Investigations	Page 227

Introduction

To help teachers teach science through investigations, Seven Sides Publishing has provided a series of lab manuals for Elementary Science, Middle School Science, Physics, Chemistry, Biology, Earth & Space Science, and Environmental Systems. These manuals are a rich resource for structure and investigations. There is a shortage of user-friendly labs that easily allow teachers and students to perform experiments quickly. Too many labs have too much busy writing within them, where teachers do not want to take the time to read everything to figure out if it would be good for them to use with their students. If the teachers do not want to read it, do you think the students do? So I have taken a lot of the traditional labs that have been around for decades and simplified them; so they are easy to read and perform. I have also added some new original labs that have never been seen before. There have been efforts to try to have teachers do more investigations with their students, but there is no plan or solution to deal with the real issues teachers have in preparing to do this. The book How to Teach Science Through Investigations has the plan, and the Simple Investigations Lab manuals have the solutions so students can learn science through investigations with minimal effort. Teaching science through investigations will make your classrooms more efficient, where students learn content and practice skills simultaneously. Science is a process of doing. Doing this process is the most efficient way for students to learn science and be able to use it in the future. We live in a culture where science-literate people are needed for jobs, but too few can be found. If you incorporate these labs with virtual labs (that I will point you to in each section of the lab manual), skill/math practice, and concept maps, you will not need to fill in gaps by giving lectures. All content can be learned through investigations and practice. Remember, we only remember 5-20% of what we hear. That 20% is when you are really interested in the content. But hearing practices no science process skills and does not activate any higher cognitive thought. Lecturing is not a good option. We remember 75-80% of what we do/experience and 90-95% of what we teach. Investigations allow us to keep our students in these higher retention percentages. Teaching through investigations also works because students spend more time in class at higher Bloom's Taxonomy levels, staying in zones C and D on the Rigor Relevance Chart when they perform investigations. And if you add the physical way they are stimulated with the hands-on experience, you cannot deny the level of learning will be much higher while students perform investigations. This manual gives you the resources you need to teach Chemistry through investigations.

We separated each of these sections in the manual like you may divide your class units. There have been many studies on presenting the order of content in efficient ways. I will be doing my best to follow this scope and sequence. I will include concept maps at the front of each section that shows the vocabulary and visual clues to how concepts relate to each other;

this has been a great way to organize information. It talks to the students to see how ideas work together, making it easy to chunk information to use at higher cognitive levels. At the beginning of each lab, I put the materials you will need in boldface in the directions; this saves time for your lab preparation. There is also a safety question in boldface just after that for you and your students to evaluate. It says, "Looking at the material and lab we will be using, what are the safety precautions we should take to protect ourselves and materials during this investigation." Make sure to read the lab so you can better answer this question with your students.

Virtual Labs

Hands-on labs are not the only way for students to learn science, but they are the most effective. However, virtual labs can be used with these hands-on labs. Many investigations physically cannot be done hands-on, so some experiments will have to be done virtually. There are three sources that I have used in the past that have a good number of resources. **Physicsclassroom.com** and **PhET.colorado.edu** are free to everyone and are great to use. **Physicsclassroom.com** has teacher notes and activities/exercises that guide students through Physics and Chemistry Interactives. You can find them under the simulation and open, download, or print the PDF. They also have a series of Concept Builders that are a tremendous virtual practice that can replace those worksheets that help students practice concepts, math, and skills. They can be hard to find, so above the list provided is the section where they can be found (underlined and in italics) on the website. **PhET.colorado.edu** has a variety of activities of different levels that you can explore to go through their simulations. They are also easy to download and print. **ExploreLearning.com** is expensive, but the quality of its product is much higher than the other two. When you click on a Gizmo, you can also click on lessons and find the Student Explorations that go with each Gizmo that you can modify, download, or print. They are written at a very high quality, making the students think like a scientist. At the end of each section of this lab manual, we will include a list of virtual labs from these organizations that would be great to use with these labs. Please remember that virtual labs should never replace hands-on labs. If the students can learn the content live, that should be the priority because it is more of an experience that will be remembered. There are many other virtual simulations out there, but none so far have moved me to use them over the three I have mentioned here.

TIPPERs

TIPPERs are great for students to explore and think about different scenarios for each concept of Physics and Chemistry. These help students think outside the box, apply concepts to real life, and think about how multiple concepts would be used together. I suggest you get the books of TIPPERs to practice and discuss after completing these labs and investigations.

Probe-ware

This lab manual has lots of labs that use probe-ware. Students must learn how to use probe-ware; this means teachers need to know how to use probe-ware. Many companies use digital probe-ware with all the research, development, testing, and forensic testing they do; this has potential career opportunities that help students become more marketable for jobs if they are familiar with using probe-ware. Hooking everything up is just as easy as charging your phone. When I was a High School Science Technology Coach and researched which companies and devices would be the most user-friendly to students, I found using Vernier Probe-ware was better for high school students, but PASCO seemed better for middle school students. Both are giants in the probe-ware industry for education. Since this Lab manual and the series were written with High School in mind (many of these labs can be used for middle school classes because they are so simple) and I am more familiar with Vernier, I will be referring to Vernier Probe-ware. However, PASCO would be a great alternative.

Interfaces are devices that the probes are connected to that talk with the program (Logger Pro) that displays the data. I found the most economical and friendliest way for students to see the data from probe-ware is to use the Vernier LabQuest Mini interface hooked up to a computer with Logger Pro. LabQuest Mini has multiple ports that are needed in many labs. They are the least expensive, so they are better on the budget. They require no batteries, so they are easy to transport if you need or want to. The other interfaces are more expensive, require batteries if you are going outside, or the stand-alone devices have a smaller screen to see the data, with less flexibility to manipulate the parameters like changing the time of data collection or changing units if you want to change or modify an experiment. Some costly wireless probes and interfaces may be easier to use if you do not mind the cost. A computer screen is much bigger and makes it easier to see the data, so this is my preferred setup. But using any interface will work fine for these labs.

Connecting the Probe-ware

To hook them up, you will plug your probe into one of the channels or the sonic on the interface. If the plug does not fit in smoothly, either you are plugging it upside-down or trying the wrong port. Then take the little chord that looks like it would go into your phone and plug that into your interface. Take the other end, and plug it into a USB port on your computer. Open up Logger Pro on your computer. If everything is hooked up properly and the computer and interface are working properly, you will see a green button at the top of the computer screen that says "Collect." Many of the labs have preset settings in Logger Pro. You will use the manila folder at the top left of the toolbar in Logger Pro to find the folders and files you will be instructed to go to for these specific settings for different labs. Whenever you get the physical equipment, they will have detailed instructions in the box they come in on how to hook them up if you are still confused. They will also have instructions on how to calibrate the probes if needed. There are a few probes that require frequent calibration. If we use any, it will be discussed in the lab directions. The more you use probe-ware, the easier it gets to set up. I usually only have to show my students twice to have them be able to set the equipment up on their own. But as you are showing them, have them physically do it. You can also find detailed instructions online at Vernier.com. Many more detailed labs can also be found there under lab ideas.

You also can use standard equipment like spring scales for force sensors or thermometers for temperature probes. Because schools want to integrate more technology, we wrote these labs to use probe-ware wherever applicable. Because they are so simple, these labs can be modified to fit whatever equipment you have. There are very few labs that I have used in my career that I did not alter how I presented them. One reason we wrote these labs this way was to customize them to the Texas TEKS. We also wrote them the way we thought teachers would want to use them.

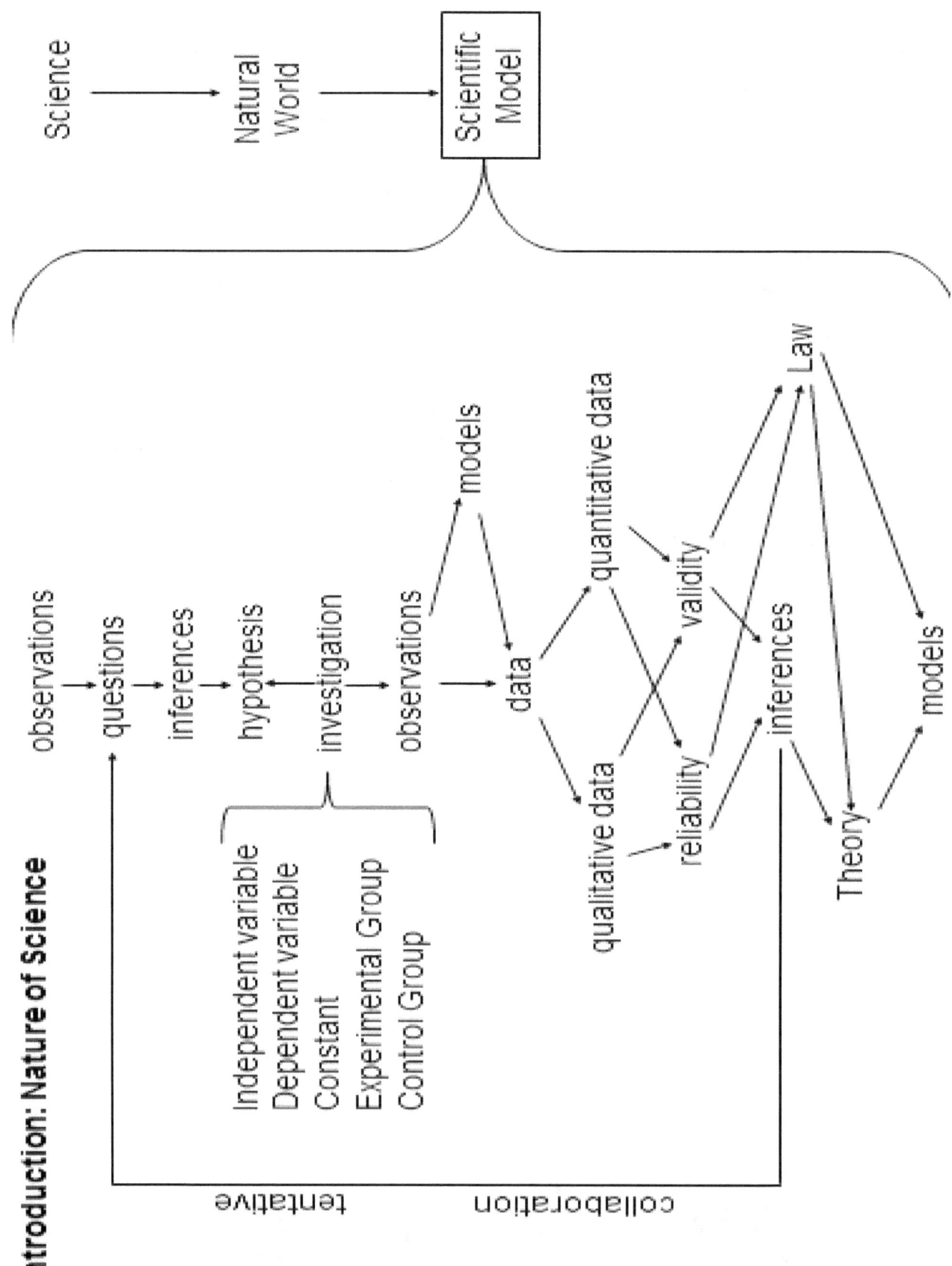

Focus on the Process

Directions:

Get a **small Legos set**. Teachers, make sure it is not too easy for your students. You are going to try to put it together in two different ways. Time how long it takes to put it together each way and answer the questions that follow. **Looking at the materials and lab we will be using, what are the safety precautions we should take to protect ourselves and materials during the investigation?**

A) Take the Lego pieces and construct the picture (the **product**) on the box's cover, looking at nothing but the cover and the Lego pieces.

B) When 20 minutes have passed, or you are done, take what you have made totally apart. Take out the directions (the **process**) and construct the product while using the step-by-step directions. Time how long it took you to complete the set.

Questions:

1) How did it feel trying to construct the Legos (A) without any directions?

2) Did you finish? If so, how long did it take?

3) How did it feel to construct the Legos (B) with the step-by-step directions?

4) Did you finish? If so, how long did it take?

5) Which strategy (A or B) allowed you to complete the product?

6) Which strategy (A or B) was more intimidating?

7) Which strategy (A or B) allowed you to see what is under the surface?

8) Which strategy (A or B) will allow you to learn more?

We often get anxious or procrastinate when faced with a large task. We are tempted to take a "shortcut" (copy or cheat, we do not learn much when we do this). There are pain and stress hormones that are released when this happens. One way to overcome this is to just worry about the next step in the process and not worry about the product. You can see and measure progress, which makes the process not feel too bad. Another way is just to start working. When you start working, those pain and stress hormones stop getting released so that anxiety goes away; this is why when we want to learn efficiently and effectively, we must:

Focus on the _____ and the _____ will take care of itself.

9) How is putting the Lego pieces together like putting ideas together to understand concepts?

Measurement Lab

Directions:

You will need **water**, a **scale**, a **meter stick**, a **temperature probe** attached to an **interface** connected to a **computer** with **Logger Pro**, a **100 mL graduated cylinder**, and a **stopwatch**. **Looking at the materials and lab we will be using, what are the safety precautions we should take to protect ourselves and materials during the investigation?**

1) Take the graduated cylinder and find its mass empty; write this in Data Table 1.
2) Add 50 mL of water to the graduated cylinder. Make sure you use the meniscus properly where the volume is at the bottom of the meniscus. Have the teacher check that you measured it correctly. Have each person in your group empty and fill the graduated cylinder with 50 mL of water. As they do so, have each person in your group time how long it takes for each person to fill the graduated cylinder and check it is correct (it is not a race, just a chance to get familiar with using the graduated cylinder and stopwatch).
3) Now find the mass of the graduated cylinder with 50 mL of water in it. Subtract the mass of the empty graduated cylinder from this mass and write the water's mass in Data Table 1.
4) Hook your temperature probe up to an interface and hook your interface up to a computer with Logger Pro (unless you have a LabQuest 2, then just hook your probe to the LabQuest 2). Find where the Logger Pro is located on your computer so you can use it again in the future. Once open, find the graduated cylinder's water temperature in Fahrenheit and Celsius (you will have to figure out how to change units). Write these in Data Table 1.
5) Take your meter stick and measure the length of the graduated cylinder. And measure the width of the base in centimeters. Write these in Data Table 1

Data Table 1

Object	Mass (g)	Volume (mL)	Time to Fill (s)	Temp (°F)	Temp (°C)	Length (cm)	Width (cm)
Graduated Cylinder		✘		✘	✘		
Water			✘			✘	✘

Questions:

1) Convert a length to meters, the volume to liters and a mass to kilograms, and Celsius to Kelvin.

   Length _____ m   Volume _____ L   Mass _____ kg   Temp _____ K

2) What do you notice about the mass of the water compared to its volume?

3) What can happen to your investigations if your measurements are not accurate or precise?

4) Why do you think the rest of the world uses the metric system over the English system.

Patterns in Pennies

Directions:

You will need a **ruler**, 10 **pennies**, a **balance**, a **roll of pennies**, and an **empty penny roll**. **Looking at the materials and lab we will be using, what are the safety precautions we should take to protect ourselves and materials during the investigation?**

1) Find the mass of one penny with a scale to the nearest .1 g. Then measure the height of the penny in millimeters. Write these in Data Table 1 below.
2) Place another penny on top of the original penny and find the mass and height of the two pennies. Write these in Data Table 1 below.
3) Keep adding pennies one by one, measuring the mass and height until you have 10 pennies on the scale.
4) Make a line graph with the mass on the (x) axis and the height on the (y) axis for the pennies on Graph 1.

Data Table 1

Number of Pennies	Mass	Height
1		
2		
3		
4		
5		
6		
7		
8		
9		
10		

Graph 1

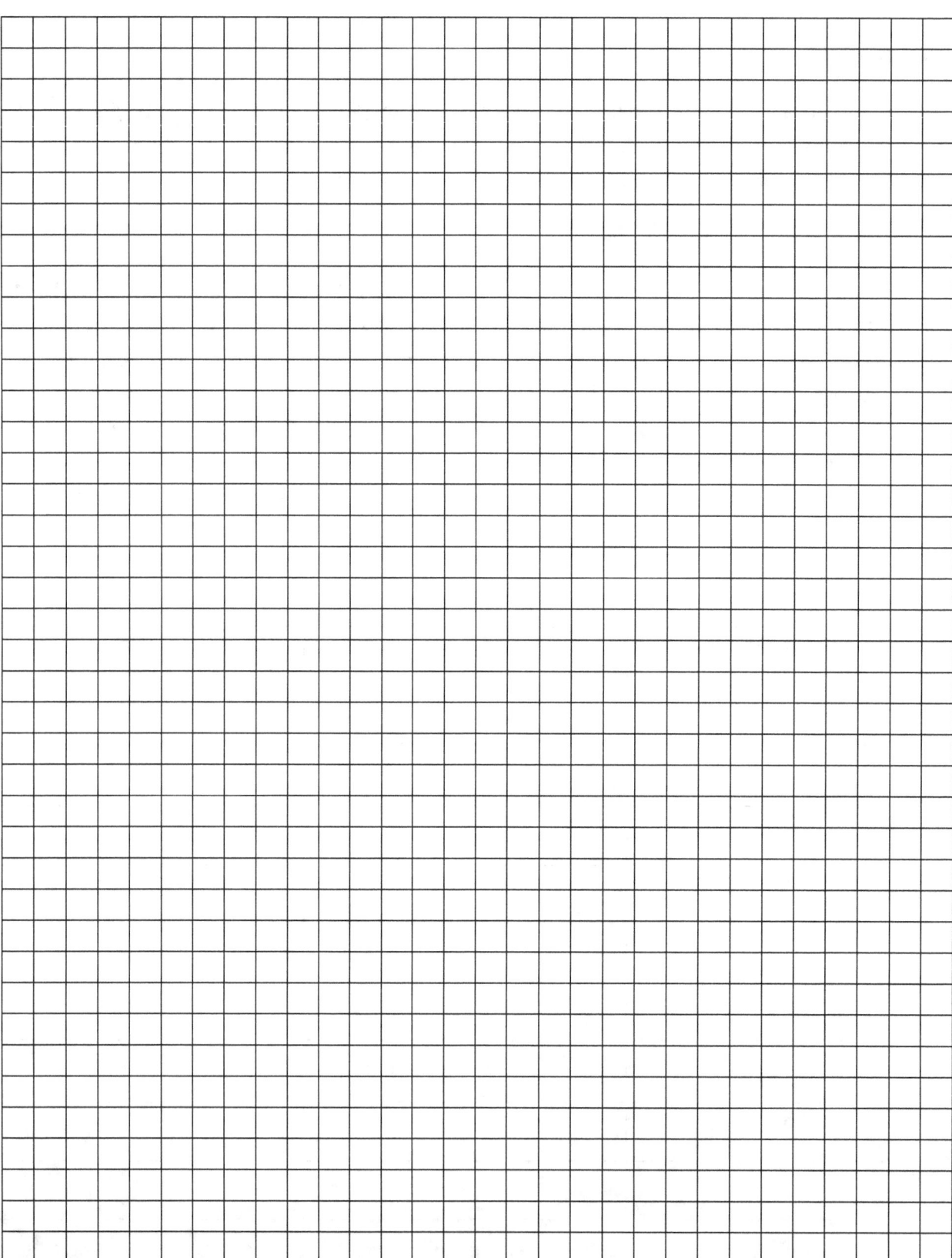

Questions:

1) What do you notice about the graph?

2) Is this a direct or inverse relationship between mass and height?

3) Do all pennies have the same mass? (Explain)

4) Do all the pennies have the same thickness? (Explain)

5) Use your data to estimate how many pennies are in the coin roll. How many pennies do you think are in the roll?

6) What did you do to estimate the number of coins?

7) What else could you do to estimate the coins?

8) Try your answer to #7. Do you get the same number as #5?

9) Carefully open up the coin roll and find out how many pennies there are. How close were you to the real number? After being done counting, carefully close the roll back up.

10) Calculate the % accuracy by taking the lowest number between your guess and the actual number dividing by the higher of the two, then multiplying by 100.

11) What were some sources of error?

Virtual Investigations that go with Introduction

ExploreLearning.com

    Unit Conversions Gizmo

    Graphing Skills Gizmo

    Measuring Volume Gizmo

    Elevator Operator (Line Graphs) Gizmo

    Weight and Mass Gizmo

    Triple Beam Balance Gizmo

    Reaction Time 1 Gizmo

    Reaction Time 2 Gizmo

Physicsclassroom.com/Concept-Builders/Chemistry:

    Measurement and Numbers

    Significant Digits and Measurements

    Metric System

    Metric Estimation

    Experiments and Variables

    Proportional Reasoning

    Calculating Slope

    Using Graphs

    Which One Doesn't Belong

Unit 1 Properties of and Changes in Matter

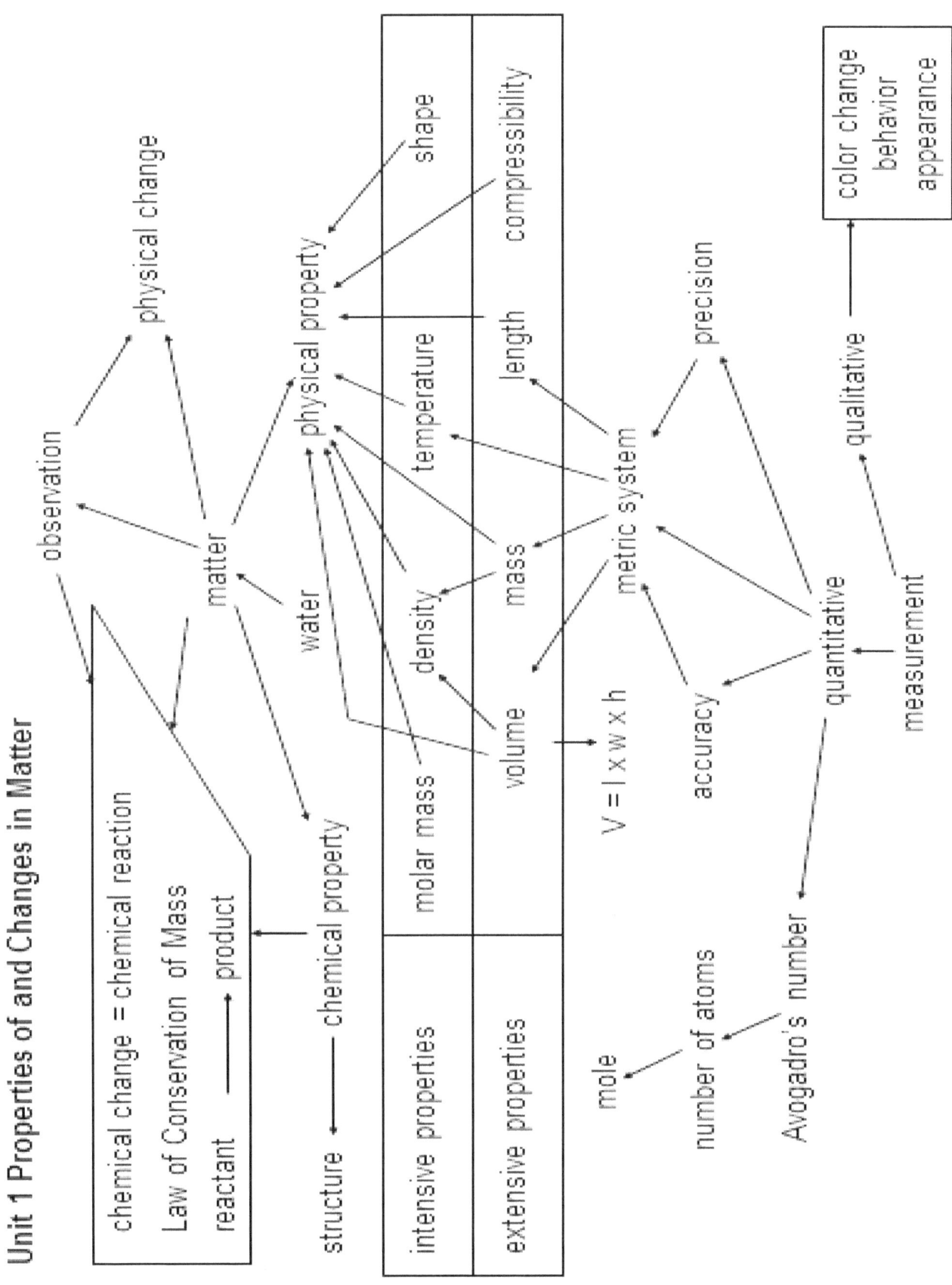

Measure and Calculate Density

Directions:

You will need a variety of **cubes** of different materials, a **ruler**, and a **scale. Looking at the materials and lab we will be using, what are the safety precautions we should take to protect ourselves and materials during the investigation?**

Fill in Data Table 1 below of the different cubes your teacher gives you by finding the mass, measuring the sides of each cube, calculating the volume, and then taking the mass divided by the volume to calculate their densities.

Data Table 1

Cube	Mass (g)	Length (cm)	Width (cm)	Height (cm)	Volume (cm^3)	Density (g/cm^3)
1						
2						
3						
4						
5						
6						
7						
8						

Questions:

1) Which object was the densest?

2) Which object was the least dense?

3) What is the relationship between mL and cm^3?

4) What is another way to find the volume of an object?

The Density of Oddly Shaped Objects

Directions:

You will need a **scale**, **water** in a **graduated cylinder**, and **any objects** that will sink in water and fit in your graduated cylinder. **Looking at the materials and lab we will be using, what are the safety precautions we should take to protect ourselves and materials during the investigation?**

1) Use a scale to find the mass of your first object. Write the name and the mass of this object in Data Table 1.
2) Fill your graduated cylinder halfway with water. Write this measurement in Data Table 1.
3) Carefully place your object in the graduated cylinder without splashing any water out. The easiest way to do this is to tip the graduated cylinder (without letting any water out) and gently slide the object into the water using the graduated cylinder's inside surface.
4) Measure the volume of the water now and place that in Data Table 1.
5) Subtract the initial volume from the final volume; this will be the object's volume. Write this in Data Table 1.
6) Now calculate the object's density by taking the mass and dividing it by the volume. Write this density on the right column of Data Table 1.
7) Repeat steps 1-6 for all your objects and write the data in Data Table 1.

Data Table 1

Object	Mass (g)	Initial Volume (mL)	Final Volume (mL)	Volume of Object (mL)	Density of Object D=m/V
					g/mL
					g/mL
					g/mL
					g/mL
					g/mL

Simple Chemistry Investigations

Questions:

1) Which object is the densest?

2) Which object is the least dense?

3) Why did the volume of water appear to rise when you put in your objects?

4) What is another way of measuring the volume of objects that are not irregularly shaped?

5) What if the object you want to measure floats in the water; how can you find its density?

6) What do you think affects the density of objects?

Which is Denser in the Mixture?

Directions and Questions:

Have a **bottle** filled with **oil** and **water. Looking at the materials and lab we will be using, what are the safety precautions we should take to protect ourselves and materials during the investigation?**

1) The liquid with the biggest density would be the liquid on the bottom, and the liquid with the smallest density would be on the top. Which liquid has the biggest density?

2) Which liquid has the smallest density?

3) Water is made up of all Hydrogen and Oxygen. Oil is made of mostly Hydrogen and Carbon and a small amount of Oxygen. Use a periodic table looking at the masses of these elements and where they sit in the periodic table, then explain why one floated on top of the other.

4) Shake up the bottle and let it sit. Do the liquids settle out the same way again? Explain why.

5) When there is an oil spill in the water, where would you expect to find the oil?

    a. How does this affect how it can be cleaned up?

Extensive and Intensive Properties

Directions:

You will need a **scale**, a **graduated cylinder**, a **ruler**, **water**, and a **temperature probe** attached to an **interface** connected to a **computer** with **Logger Pro**. Looking at the materials and lab we will be using, what are the safety precautions we should take to protect ourselves and materials during the investigation?

1) <u>**Extensive properties**</u> tell you how much of a substance you have. <u>**Intensive properties**</u> are properties of a substance that does not change no matter how much of the substance you have. We will take measurements and classify those measurements as extensive or intensive.
2) Take your empty graduated cylinder and find its mass with the scale. Put this measurement in Data Table 1
3) Now put 75 mL of water in it. Find the mass now and put that in Data Table 1.
4) Subtract the mass of the empty graduated cylinder from the graduated cylinder with 75 mL in it. Write this as the mass of 75 mL of water in Data Table 1.
5) Now take your ruler and measure the height of the water in the graduated cylinder. Write this in Data Table 1.
6) Measure the temperature of the water in the graduated cylinder. Write this measurement as the temperature for 75 mL of water in Data Table 1.
7) Now add 25 mL of water to the graduated cylinder giving it 100 mL. What are the masses now for 100 mL of water? Write this in Data Table 1.
8) Measure the temperature of the 100 mL of water and write that in Data Table 1.
9) Take the ruler and measure the height of the water 100 mL of water in the graduated Cylinder. Write this in Data Table 1.
10) Now calculate the water density by dividing its mass by its volume for 75 mL and 100 mL of water.
11) Using the information in Data Table 1 and the answers to the questions, decide whether the measurements for **mass, volume, temperature, height**, and **density** are extensive or intensive and write in Data Table 2. The property is extensive if the numbers went up because we added more substance. If more substances did not cause the measurement to go up, then the property is intensive.

Data Table 1

Property	Empty Graduated Cylinder	Graduated Cylinder and 75 mL H2O	75 mL of H2O by its self	Graduated Cylinder & 100 mL H2O	100 mL of H2O by its self
Mass					
Volume	X	X	75mL	X	100mL
Temperature	X	X		X	
Height	X	X		X	
Density	X	X		X	

Questions:

1) Did the volume change?

2) Did the temperature change? If it did, did it change much, or did the amount of water cause the change?

3) Did the height of the water change when more was added?

4) Did the mass change when more water was added?

5) Did the density change when more water was added?

Data Table 2

Extensive Properties	Intensive Properties

Physical and Chemical Changes

Directions and Observations:

You will need **safety goggles**, an **aluminum pan**, **paper**, a **pipette**, a **candle**, a **lighter** or **matches**, **salt**, a **beaker of water**, **vinegar**, **baking soda** in a **small beaker**, **steel wool**, **long forceps**, a **clean piece of metal** and **another that is corroded** (of the same kind of metal), **hydrogen peroxide**, and **liver** or **banana**. **Looking at the materials and lab we will be using, what are the safety precautions we should take to protect ourselves and materials during the investigation?**

<u>**Physical changes**</u> are changes that happen to a substance that does not cause any new substances to form. <u>**Chemical changes**</u> are changes that occur when new substances are formed.

1) Make sure everyone is wearing their protective goggles. Have your teacher come by and light the candle. What do you see happening to the wick of the candle?

2) Is this creating any new substances? Fill in Data Table 1 for the burning wick.

3) What do you see happening to the wax of the candle?

    a. Is this creating any new substances?

    b. Fill in Data Table 1 for the melting wax.
4) Take a piece of paper and tear it up. What did you see happen to the paper?

5) Did tearing the paper create any new substances? Fill in Data Table 1 for tearing paper.

6) Wad up a piece of paper and put it in the aluminum pan. Make sure everyone is wearing their protective goggles. Have your teacher come by with the lighter and light the paper on fire. Make sure to keep everything away from the flames. What do you see happening to the paper?

   a. Is it creating any new substances? Fill in Data Table 1 for burning paper.

7) Pour the salt into a beaker of water and stir it up. What do you see?

   a. Did this create any new substances? Fill in Data Table 1 for dissolving salt.

8) Now pour some vinegar into the small beaker of baking soda. What happened?

   a. Did this make any new substances? Fill in Data Table 1 for mixing vinegar and baking soda.

9) Look at the two pieces of metal. What difference do you see between the clean metal and the corroded?

10) Did corrosion cause a change to create a new substance? Fill in Data Table 1 for the metal.

11) Make sure you are wearing your safety goggles. Look at your steel wool and hold it with your forceps. Now have your teacher expose it to fire with the lighter. What do you see happen? Look at the color when your teacher is done (only the part of it that was lit).

   a. Did this produce any new substances? Fill in Data Table 1 for burning steel wool.

12) Finally, take some hydrogen peroxide and pour it on some liver or a slice of banana. What do you see happen?

   a. Did this produce a new substance? Fill in Data Table 1 below for hydrogen peroxide on life.

Data Table 1

Change that Happened	Physical or Chemical Change?	Evidence Why
Burning wick		
Melting wax		
Tearing Paper		
Burning paper		
Dissolving salt		
Vinegar and baking soda		
Corrosion of metal		
Burning of steel wool		
Hydrogen peroxide on life		

Questions:

1) Describe observations you might see when a physical change occurs.

2) Describe observations you might see when a chemical change occurs.

3) How could you tell dissolving sugar in water is a physical change?

4) Why does bubbling let you know there is a chemical change?

5) How do you know burning something is a chemical change?

Virtual Investigations that go with Properties and Changes in Matter

ExploreLearning.com:

    Density Laboratory Gizmo

    Density via Comparison Gizmo

    Density Gizmo

    Density Experiment: Slice and Dice Gizmo

    Determining Density via Water Displacement Gizmo

    Archimedes' Principle Gizmo

    Chemical Changes Gizmo

    Mystery Powder Gizmo

    Moles Gizmo

Phet.colorado.edu:

    Density

Physicsclassroom.com/Concept-Builders/Chemistry:

    Classification of Matter

    Chemical vs. Physical Properties

    Density Ranking Tasks

    Mole Conversions

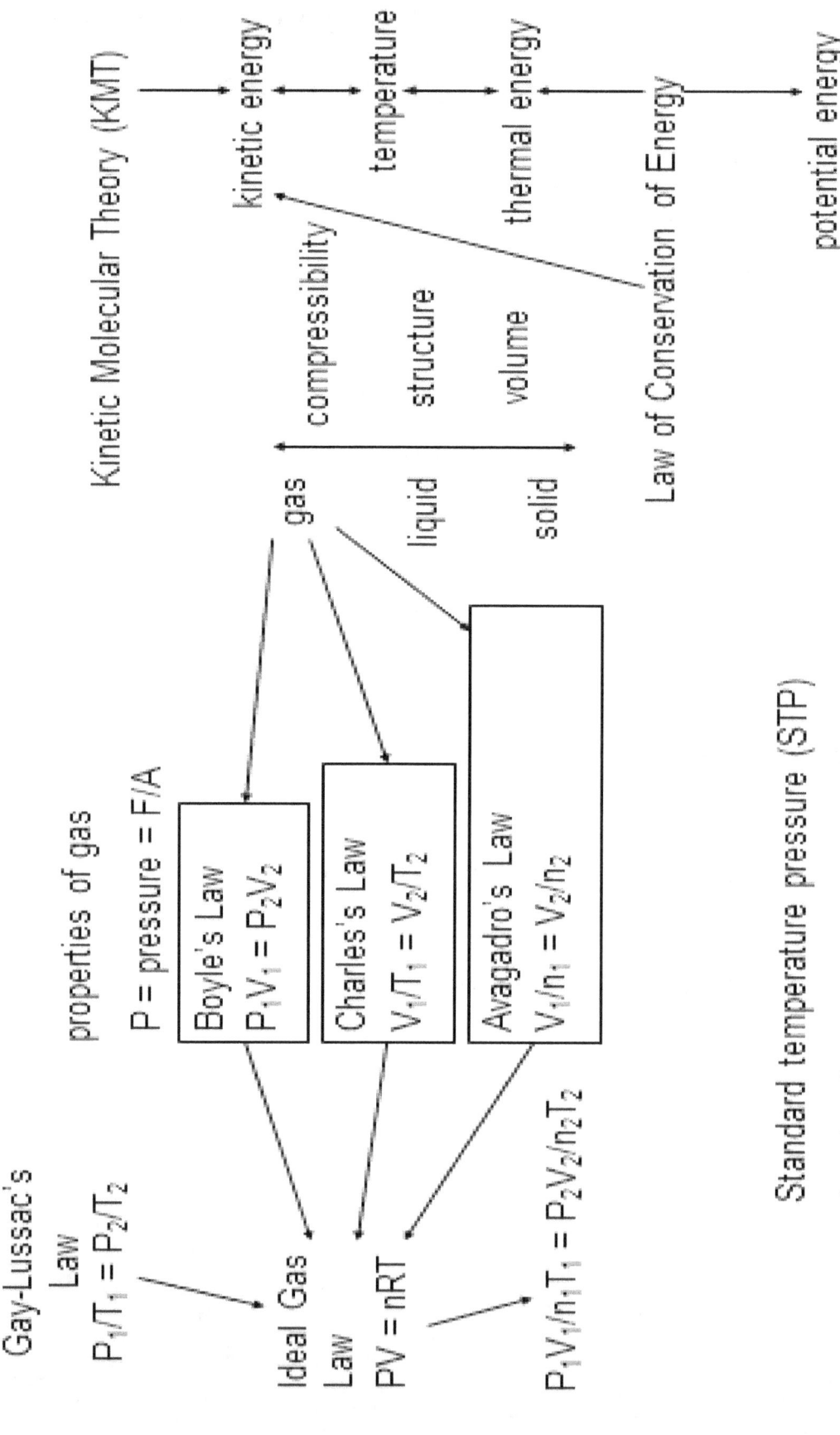

Observing Molecular Motion

Directions:

You will need at least three of the **biggest beakers** in your school. They all will be filled with water. One beaker **filled with water** needs to be put in a **refrigerator** so the water is cold. You will also need one beaker on a **hotplate** before doing the demo, so it heats up (if it boils, you can turn the heat off). The third beaker will be room temperature water straight out of the tap. Lastly, you will need some **food coloring. Looking at the materials and lab we will be using, what are the safety precautions we should take to protect ourselves and materials during the investigation?**

1) Line all three beakers up from coldest to warmest where the whole class can see. Place a drop of food coloring in the cold beaker and have the students watch how the food coloring spreads.
2) Put a drop of food coloring in the room temperature water, then in hot water on the hotplate. Have students observe the movement in all three beakers. It will not take long for the hot one to become homogeneous.
3) Have the students draw what they see in the three beakers below.

       Cold   Warm   Hot

Questions:

1) Which beaker had the dye move the fastest?

2) Which beaker had the dye move the slowest?

3) Why do you think this happened this way?

4) Temperature is defined as the average kinetic energy of molecules. How does this explain what was happening in the hot, warm, and cold water?

5) Which beaker had the most energy?

6) How can this explain why we get hurt when we touch something hot?

Simple Chemistry Investigations — Seven Sides Publishing

Observing Boyles Law

Directions and Questions:

You will need a **large syringe**, a **small marshmallow**, another **large syringe** but with a **stopcock**, an **industrial suction cup**, a **plastic bottle** with a **valve** fixed to the **cap**, an **air pump**, and a **small sealed syringe. Looking at the materials and lab we will be using, what are the safety precautions we should take to protect ourselves and materials during the investigation?**

1) You will need a large syringe and a small marshmallow.
    a. Put a small marshmallow in the syringe. With a large volume in the syringe, put your finger over the opening to block air and squeeze the syringe. What happened to the marshmallow?

    b. What caused this?

    c. Now take your finger off the opening letting the air out. Move the syringe to a low volume, put your finger back over the opening and increase the volume. What happened to the marshmallow?

    d. What caused this?

2) You will need a large syringe and stopcock.
    a. Put the stopcock on the syringe with a large volume in the syringe. Press the syringe making the volume inside smaller. What do you notice about how the pressure is felt with respect to volume?

3) You will need industrial suction cups.

a. Put a suction cup on a very smooth surface. Pull the lever on the cup to increase the volume inside the cup on the surface. Pull up on the cup. What do you notice?

b. Why do you think this happened?

4) You will need a plastic bottle with a sealed syringe inside that has a valve fixed to the cap to pump air into the bottle from an air pump. Make sure that the syringe inside the bottle has at least ½ to ¾ of the volume open.
   a. Pump air inside the bottle with the syringe inside. What do you notice about the volume of the syringe?

   b. Why do you think that happened?

   c. Now let the air out of the bottle and watch what happens to the volume on the syringe. How did the volume change?

   d. Why do you think this happened?

5) How do volume and pressure affect each other in a sealed system?

6) How is Boyles Law a law?

   a. Why is it not a hypothesis or theory?

Relationship between Temperature Volume and Pressure: Charles Law and Gay-Lussac's Law

Equipment and Safety:

You will need a **balloon**, **string**, a **meter stick**, a **freezer**, an **Erlenmeyer flask**, a **rubber stopper assembly**, a **gas pressure sensor**, **plastic tubing**, a **temperature probe** attached to an **interface** that is connected to a **computer** with **Logger Pro**, two **ring stands** and **clamps**, a **beaker with cold water**, and a **hotplate**. Looking at the materials and lab we will be using, what are the safety precautions we should take to protect ourselves and materials during the investigation?

Charles Law: the relationship between temperature and volume in a sealed flexible system.

1) At the beginning of the lab, blow up a balloon, measure its circumference with some string and measure that length with a meter stick. Then put the balloon into a freezer. Come back and check the balloon after the lab.

   a. Circumference before freezer:

   b. Circumference after freezer:

   c. What is the relationship between temperature and volume in a sealed flexible system where pressure does not change?

Gay-Lussac's Law: the relationship between temperature and pressure in a sealed system.

2) You will need an Erlenmeyer flask, a rubber stopper assembly that you can seal the flask up with, and connect a gas pressure sensor to it with plastic tubing. You will also need a temperature probe. You will attach the gas pressure sensor and temperature probe to an interface connected to a computer with Logger Pro. You will need a clamp connected to a ring stand to hold the sealed Erlenmeyer flask inside a beaker of cold water on a

hotplate. You will need another clamp and ring stand to keep the temperature probe suspended in the water, not touching any glass during the experiment.

    a. In Logger Pro, open the Chemistry with Vernier folder and open file #07 Pressure-Temperature. Make sure everything is set up and secure. Press "Collect."

    b. Turn on the hotplate.

    c. Once you see the water is about to boil (when the temperature is close to 100°C), click "Stop" and turn off the hotplate.

    d. In Logger Pro, look at the temperature vs. pressure graph. What is the relationship between temperature and pressure if the volume cannot change?

    e. Try and adjust the graph so you can see the temperature down to -273°C. What would the pressure be?

    f. Find out the volume of any gas at -273°C (absolute zero). Look it up on the internet.

    g. -273°C is absolute zero where no energy exists. What implications does this have in light of your finding?

    h. What do you think the temperature might be at the bottom of a black hole?

3) How are the relationships between temperature, volume, and pressure explained as laws?

    a. Why are they not theories?

Ideal Gas Law Calculations

Directions and Questions:

Use the **Ideal Gas Law equation** and **Dalton's Law of Partial Pressure** to answer the following questions.

1) What volume is occupied by 100 g of O_2 at a pressure of 1.75 atm at a temperature of 34°C?

2) If a 2-liter bottle can hold 3 moles of gas and is heated to 500°C, what is the can's pressure?

3) If a 2-liter bottle can hold 7 moles of gas heated to 500°C, what is the can's pressure?

4) If a 2-liter bottle can hold 10 moles of gas heated to 500°C, what is the can's pressure?

5) If we take the three cans from problems 2, 3, and 4 and put all their gases in one of the containers, what would be a new pressure?

6) If a gas in a 75 mL container at 30°C has 2.5 moles in it, what is its pressure?

7) If 2.5 atm of pressure is in a 25 mL container at 30°C, how many moles are in there?

8) If 4 moles of gas at 1.5 atm of pressure are in a container at 30°C, what is the volume?

9) If all three gasses from problems 6, 7, and 8 were put into a 100 mL container, what would be the total pressure?

10) Using problems #1 and #8, if the temperature is -273°C, what is the gas volume?

   a. What is the volume for any gas at -273°C?

   b. What special name do we give this temperature?

   c. Why do you think the volume is this number?

Virtual Investigations that go with Gas Laws

ExploreLearning.com:

    Temperature and Particle Motion Gizmo

    Boyle's Law and Charles's Law Gizmo

    Ideal Gas Law Gizmo

    Equilibrium and Pressure Gizmo

    Diffusion Gizmo

Phet.colorado.edu:

    Diffusion

    Gas Properties

    Gases Intro

    Balloons and Buoyancy

    States of Matter

Physicsclassroom.com/Concept-Builders/Chemistry:

    Pressure Concepts

    Pressure and Temperature

    Volume and Temperature

    Pressure and Volume

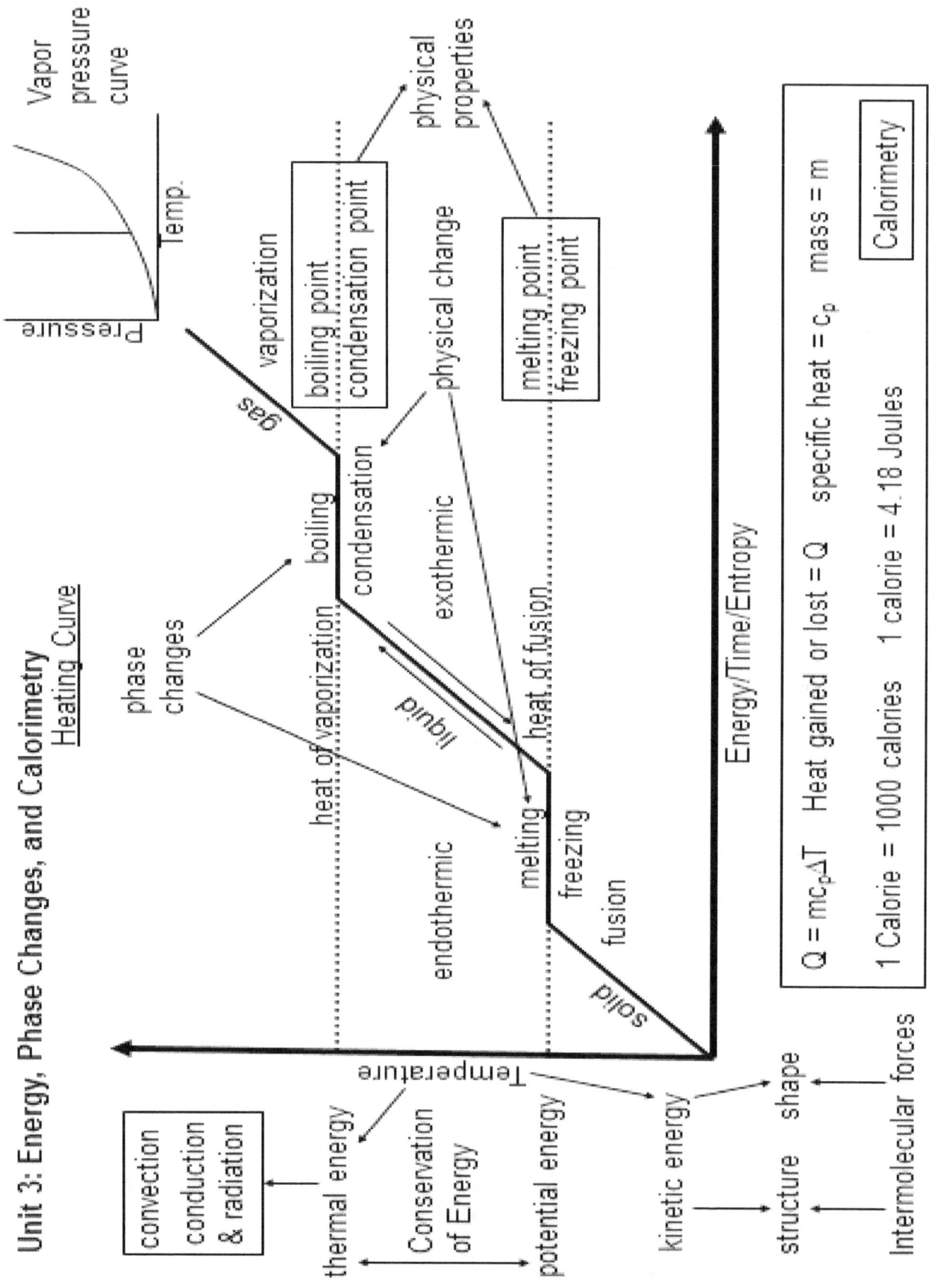

The Four Laws of Thermodynamics

Directions:

Use the **internet** and your **textbook** to research the Four Laws of Thermodynamics. Explain them and give applications and examples of how each are applied in everyday life.

1) **0th Law:**

   Applications & Examples:

2) **1st Law:**

   Applications & Examples:

3) **2nd Law:**

   Applications & Examples:

4) **3rd Law:**

   Applications & Examples:

Student Atomic Motion

Directions and Questions:

You will need a class of **students** (that will act like atoms as a solid, liquid, and gas). **Looking at the materials and lab we will be using, what are the safety precautions we should take to protect ourselves and materials during the investigation?**

1) Have students sit in their seats. They are now modeling how atoms move when they are in a **solid-state**. Are they absent of any motion? If not, describe how the student molecules are moving.

2) Now have the students get up and walk slowly around in a small area of the class; they are now modeling **liquid** atoms. How is this motion of the **liquid state** different from the solid state?

3) Now have the students walk around the classroom faster over the entire classroom. When they are about to bump into another person, have them not touch; they just move quickly in another direction. The students are now modeling how atoms move in a **gaseous state**. How do the **gas** atoms move differently from the other two states?

4) Which state of matter had the most energy moving in it? Explain how you can tell.

5) Which state of matter had the least energy moving in it? Explain how you can tell.

6) How could you measure the density of the molecules in this activity?

7) Which state of matter had the smallest density?

8) Which state of matter had the largest density?

9) Which states of matter could change shape to fill the container?

10) Which state of matter could not change shape?

11) Which state of matter was able to fill the whole container?

12) How could we change this model to show molecules (atoms bonded together) in motion as solid, liquid, and gas?

13) How was this model not accurate in showing atomic motion?

Seeing the Heating Curve

Directions:

You will need a **beaker**, **frozen water**, and a **temperature probe** suspended in it; this needs to be prepared for the day before in the **freezer**. Freezing water in the beaker many times breaks the beakers, so freeze the water in paper or **Styrofoam cups** while suspending the temperature probes in the water. When doing the investigation, peel off the cup and put the ice in the beaker. Attach the probe to an **interface** that is connected to a **computer** with **Logger Pro**. You will need to place the beaker of ice on a **hotplate**. You also will need a **ring stand** and **clamp** to hold the temperature probe up as the ice melts. **Looking at the materials and lab we will be using, what are the safety precautions we should take to protect ourselves and materials during the investigation?**

1) Make sure the Logger Pro is set up to collect temperature data every second for at least 20 minutes.
2) Click "Collect" and turn the hotplate on high.
3) Watch the setup and data until the ice melts to water, and then the water boils for a little while. Then click "Stop."
4) Use the graph the Logger Pro gives you to answer the questions below.

Questions:

1) What was the temperature range of the ice in the data? Change in temperature shows kinetic energy is changing for water.

2) What did the graph look like while the water was ice (before it started to melt)? As the temperature changed, so did the kinetic energy.

3) What does the graph look like as the ice melted (changed from solid to liquid)? This pattern shows potential energy is changing.

4) Is the kinetic energy changing during this phase change?

5) What was the temperature range of the water after the ice melted and before it boiled? This temperature change is the change in kinetic energy.

6) As the water boiled, what was the temperature?

7) What does the graph look like during this phase change? Potential energy is changing again, but the kinetic energy is not changing.

8) What do you think the temperature range would be for steam?

9) Build and label Graph 1 of what the graph looked like on the Logger Pro.

Graph 1

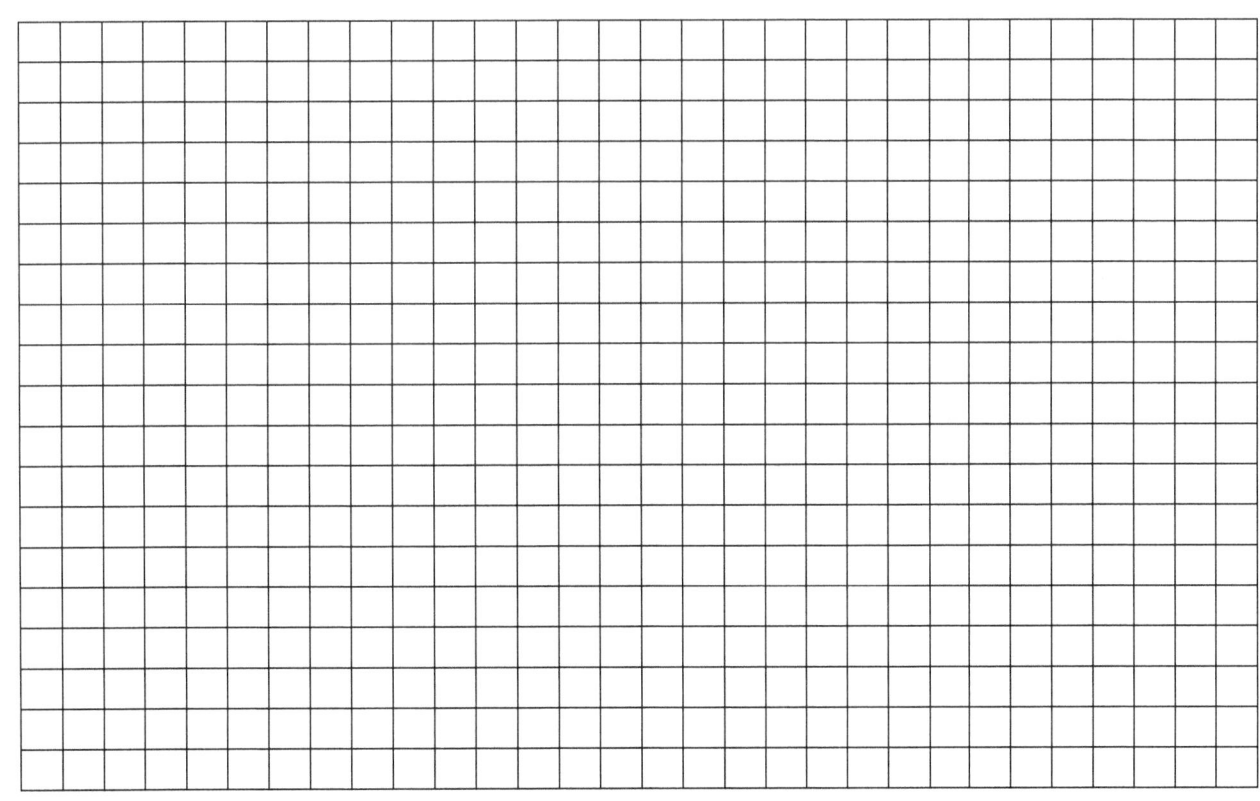

Time (min)

Calorimetry Lab

Directions:

You will need **safety goggles**, a **scale**, a **lighter**, a **paper clip**, a **peanut**, a **small beaker**, **water**, a **ring stand**, two **clamps**, and a **temperature probe** attached to an **interface** connected to a **computer** with **Logger Pro**. **Looking at the materials and lab we will be using, what are the safety precautions we should take to protect ourselves and materials during the investigation?**

1) Take the paperclip and bend it to make a stand for the peanut to sit on.
2) Measure the mass of the peanut and write it in Data Table 1. Then set the peanut on the paperclip stand you created.
3) Measure the mass of the small beaker (the smaller, the better). Write the mass in Data Table 1 below.
4) Fill the beaker no more than halfway full and measure the mass of the beaker and water. Then subtract the mass of the beaker to get the mass of just the water. Write all this in Data Table 1.
5) Set up the ring stand with a clamp that will hold the small beaker. Make sure to set the beaker directly over the peanut.
6) Have the second clamp hold the temperature probe to where its tip is suspended in the water. Set the Logger Pro to collect the temperature at °C for 5 minutes. Write down the initial temperature of the water in Data Table 1. Click "Collect" and make sure your safety goggles are on.
7) Take the lighter and light the peanut on fire. Watch the peanut and the temperature as the peanut burns. When it is done burning, look at the temperature in °C. When the temperature stops moving up and starts to go down, click "Stop" to stop collecting data. Take the highest temperature and write that down as the final temperature in Data Table 1.

Data Table 1

Mass peanut	Mass of Beaker	Mass of Beaker & H2O	Mass of H2O	Initial Temp of H2O	Final Temp of H20
g	g	g	g	°C	°C

Questions:

1) Use the formula for specific heat $Q = m \times c_p \times \Delta T$ to calculate how much energy was absorbed by the water. How much energy did the water absorb from the peanut?

2) Convert that to calories to show how many calories were absorbed by the water.

3) Was this all the energy that the peanut gave off?

4) What we made was a very primitive calorimeter to measure the number of calories a food sample contains. Why was our calorimeter not very accurate or precise?

5) How could we improve the calorimeter?

Extra Credit: If you have a **calorimeter**, set up the experiment again using the actual calorimeter. The only thing you would need to change on your data table is to write the word calorimeter in for the beaker. The peanut will sit in the bottom container. Try and calculate how many calories are in a peanut.

Simple Chemistry Investigations

Observing and Calculating Change in Energy

Directions:

You will need some **safety goggles**, an **apron**, a **graduated cylinder**, **ammonium nitrate** $NH_4NO_3(s)$ or **urea** $(NH_2)_2CO(s)$, **Sodium hydroxide NaOH(s)**, a **scoopula** to handle the solid reactants (*do not touch them*), **water**, a **400 mL beaker**, two **Styrofoam cups**, and a **temperature probe** attached to an **interface** connected to a **computer** with **Logger Pro.** Set up to collect data for 500 seconds. **Looking at the materials and lab we will be using, what are the safety precautions we should take to protect ourselves and materials during the investigation?**

Part 1

1) Put a Styrofoam cup into the 400 mL beaker. Add 50 mL of water into the cup. Place the temperature probe into the cup with water.
2) Obtain 10 g of ammonium nitrate or urea. Gently stir the water with the temperature probe until the temperature remains stable.
3) Begin collecting data for 15 seconds. Then add the urea or ammonium nitrate. Keep gently stirring. Stop the data collection after the temperature bottoms out and begins to rise.
4) Use the statistics (STAT) button to find the maximum and minimum temperatures. Subtract the two temperatures; this is your ΔT. What is it?

5) Now use the formula for specific heat: $Q = m \times c_p \times \Delta T$ to calculate the amount of energy absorbed by the reaction ($c_p = 4.18 J/g°C$). How much energy was absorbed?

6) Dispose of the material as directed by your teacher and rinse your temperature probe and graduated cylinder.

Part 2

7) Place a new Styrofoam cup into your 400 mL beaker. Add 100 mL of water to the cup. Place the temperature probe into the water.
8) Obtain 2 grams of NaOH(s). Gently stir the water with the temperature probe until the temperature remains stable.
9) Begin collecting data for 15 seconds. Then add the sodium hydroxide. Keep gently stirring. After the temperature maxes out and begins to drop, stop the data collection.
10) Use the statistics (STAT) button to find the maximum and minimum temperatures. Subtract the two temperatures; this is your ΔT. What is it?

11) Now use the formula for specific heat: Q = m x c_p x ΔT to calculate the amount of energy given off by the reaction (c_p = 4.18J/g°C). How much energy was given off?

12) Dispose of the material as directed by your teacher and rinse your temperature probe and graduated cylinder.

Questions:

1) Exothermic reactions give off heat; endothermic absorb heat. Which reaction was exothermic, and which was endothermic?

2) How could we use reactions like this to apply to our daily lives?

Convection in Liquids and Gases

Directions and Observations:

Fill a **large beaker** with **water**, add **pepper** to it, place it on a **hotplate,** and heat the water to just below the boiling point with your **safety goggles** on. **Looking at the materials and lab we will be using, what are the safety precautions we should take to protect ourselves and materials during the investigation?**

1) Draw a picture of the motion of the pepper in hot water:

2) Describe the motion you see in hot water:

3) What do you think is causing this motion (go into detail)?

4) Light a **candle** with a **match/lighter** and gently blow it out. Which direction does the smoke go?

Questions:

1) Describe how the particle of the pepper moved as the water became hotter.

2) Describe how the particles of pepper moved as the water became colder after losing heat on the surface.

3) Explain how convection currents formed in the beaker.

4) Explain why the motion of the particles changed as the burner heated up the water.

5) Which direction did the smoke go when the candle was blown out?

6) Explain why the smoke went in that direction.

Observing Conduction Convection and Radiation

Directions and Questions:

You will need **safety goggles**, a **hotplate**, **Jiffy Pop popcorn**, a **hot air popper**, **unpopped popcorn**, **microwave popcorn**, and a **microwave**. **Looking at the materials and lab we will be using, what are the safety precautions we should take to protect ourselves and materials during the investigation?**

1) Heat popcorn in the Jiffy Pop skillet on a hotplate. Once the popcorn gets hot, what happens to it?

2) Describe how the energy moved from the hotplate to the popcorn in as much detail as you can.

3) How do you know the energy got there?

4) What method of heat transfer was this?

5) Put popcorn in a running hot air popcorn popper. What happens when the popcorn gets hot?

6) Describe how the energy moved from the popper to the popcorn in as much detail as you can.

7) How did you know the energy got there?

8) What method of heat transfer was this?

9) Microwave a bag of popcorn. What happens to the popcorn?

10) Describe how the energy got to the popcorn in as much detail as you can.

11) How do you know the energy got there?

12) What method of heat transfer was this?

Simple Chemistry Investigations — Seven Sides Publishing

Energy Transformation Balls

Directions:

You will need a pair of **steel energy transformation balls**, an **index card** or piece of **paper**, and **safety goggles. Looking at the materials and lab we will be using, what are the safety precautions we should take to protect ourselves and materials during the investigation?**

1) Put on your safety goggles. Have one person take the steel energy transformation balls and hold one in each hand. Have another person vertically hold out a piece of paper or an index card. The person holding the energy transformation balls should then smash the two balls together on the paper with a very strong force. Make sure not to get anyone's fingers in the way.

2) Observe what happens to the paper where the balls hit and what you smell.

3) Discuss with your teacher the questions that follow.

Questions:

1) What do you see on the paper where the balls hit?

2) What do you smell?

3) What do you think happened?

4) How was energy transformed?

5) How is this collision like a meteor hitting the Earth?

6) How does this relate to the theory of relativity E = mc²?

7) How does this relate to the Big Bang Theory?

8) How does this relate to The Law of Conservation of Energy?

9) Is this showing conduction convection or radiation? Explain.

Virtual Investigations that go with Energy Phase Changes and Calorimetry

ExploreLearning.com:

    Phases of Water Gizmo

    Phase Changes Gizmo

    Calorimetry Lab Gizmo

    Feel the Heat Gizmo

    Conduction and Convection Gizmo

    Radiation Gizmo

    Heat Transfer by Conduction Gizmo

Phet.colorado.edu:

    Energy Forms and Changes

    States of Matter

    States of Matter: Basics

Unit 4: Pure Substances & Mixtures

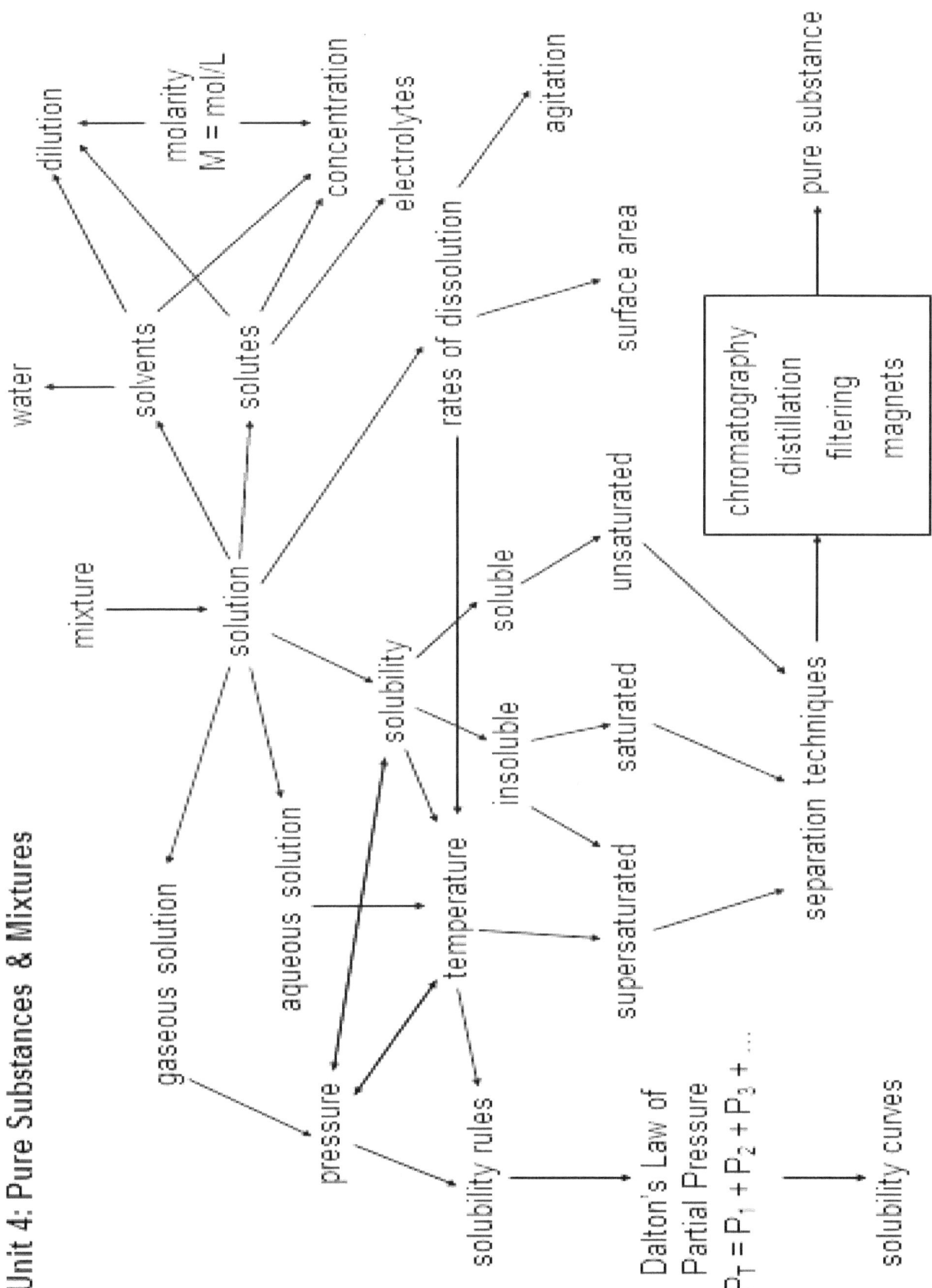

Speed of Dissolving Solutes Lab

Directions and Questions:

You will need **safety goggles**, six **beakers**, cold and warm **water** (cold water could be cooled in the **fridge** overnight or between classes, warm should be heated on a **hotplate**), six **sugar cubes**, a **plastic baggy**, and **beaker tongs**. **Looking at the materials and lab we will be using, what are the safety precautions we should take to protect ourselves and materials during the investigation?**

1) **Surface Area:** Crush one sugar cube in a plastic baggy. Heat two beakers with equal amounts of water on a hotplate. Just before they boil, drop in the crushed sugar cube in one beaker and a full sugar cube in the other at the same time. Which one dissolves the fastest?

2) **Stirring or not:** Take two beakers with equal amounts of water and place a sugar cube in each at the same time. Stir one beaker and leave the other alone. Which beaker dissolves the sugar cube first?

3) **Hot or Cold:** Take one beaker and heat it on a hotplate. Just before it boils, take a beaker of water out of the fridge. Drop a sugar cube in each at the same time. Which beaker dissolves the sugar cube first?

4) How do you dissolve a substance the fastest way possible?

Heat and Saturating Solutions

Directions:

You will need **safety goggles**, a **test tube**, **test tube tongs**, a **test tube rack** (2 or 3 for the whole class), a **hotplate**, a medium-size **beaker of water** that has room for test tubes to sit in safely, **sugar**, and a **solubility curve** for a variety of substances. **Looking at the materials and lab we will be using, what are the safety precautions we should take to protect ourselves and materials during the investigation?**

1) Heat a beaker of water on a hotplate. Heat it but do not let it boil violently.
2) Put some water in a test tube. Stir in sugar until it all dissolves. Then stir in more sugar until no more will dissolve.
3) Place test tube tongs on your test tube and place the test tube in the beaker of hot water. Stir the sugar in the tube; does it dissolve?

4) Put more sugar in the test tube, stirring by holding the tongs and wiggling the test tube back and forth. See that it dissolves. Do this a few times, then place the test tube in a test tube rack and put it into the refrigerator. Check on this at the end of the period or the next day.

Questions:

1) Which water will dissolve more sugar in it, hot or cold?

2) How could you tell?

3) What happened to the sugar solutions you put in the fridge? What you saw was a supersaturated solution.

4) What do you think will happen to the amount of a solute that will dissolve into solvent/water as the temperature of the solvent/water goes up?

5) What do you think will happen to the amount of solute that can dissolve into solvent/water as the temperature of the solvent/water goes down?

6) Find and look at a solubility curve for various substances dissolving in water. Were you correct?

7) What do you think will happen to the amount of solute that can be dissolved in a solvent/water if you double the amount of solvent/water?

Sugar or Salt

Directions:

You will need a **graduated cylinder**, two **beakers**, **salt**, **sugar**, **water**, **stirring rods**, and a **scale**. **Looking at the materials and lab we will be using, what are the safety precautions we should take to protect ourselves and materials during the investigation?**

Part 1:

1) Put 100 mL of water in each beaker. Find the mass of the first beaker with the water. Write that mass in Data Table 1.
2) Add salt to the beaker until no more salt will dissolve. Write the mass of the beaker water and salt in Data Table 1.
3) Subtract the water and beaker from this measurement, and the result is the salt's mass. Write this in Data Table 1.
4) Repeat the procedure for # 1-3, doing the same for sugar.

Data Table 1:

Measurements	Salt	Sugar
Mass of beaker and 100mL water (g)		
Mass of beaker, water, and solute (g)		
Mass of solute (g)		

Questions:

1) Which is more soluble at room temperature, sugar or salt?

2) How much difference is there?

3) How much salt and sugar do you think can dissolve in 10 g of water?

Directions Part 2:

1) Test it using the same procedures above but with only 10 mL of water in each beaker. Write your data in Data Table 2.

Data Table 2

Measurements	Salt	Sugar
Mass of the beaker and 10mL of water (g)		
Mass of the beaker, water, and solute (g)		
Mass of the solute (g)		

Questions:

1) How does this compare to your prediction?

2) When we add solvent, how does that affect the amount of solute that can be dissolved?

3) Calculate the Molarity for all four solutions made in the investigation.

Boiling Points of Solutions

Directions:

You will need **safety goggles**, a **hotplate**, two **beakers** of **water**, **salt**, a **ring stand**, two **clamps**, and two **temperature probes** attached to an **interface** connected to a **computer** with **Logger Pro**. **Looking at the materials and lab we will be using, what are the safety precautions we should take to protect ourselves and materials during the investigation?**

1) Dissolve a large amount of salt in one beaker of water but none in the other. Put both beakers on a hotplate.
2) Suspend the two temperature probes in the middle of the water in both beakers with the clamps and ring stand.
3) Turn on the hotplates and wait for the water to boil. Which beaker do you think will have the highest boiling point and which will have the lowest?

4) When the beakers reach boiling point, write their temperatures in the data table below.

Data Table:

Boiling Point of Water Only	Boiling Point of Salt Solution

Questions:

1) Which beaker had the highest boiling point?

2) What effect did dissolving a solute have on the boiling point?

3) What effect do you think it will have on the freezing point?

Making Ice Cream

Directions:

You will need **gloves**, **milk**, a **spoon**, **crushed ice**, **salt**, **Nesquik** (chocolate or strawberry), a **gallon-size Ziplock bag**, a **quart-size Ziplock bag**, and a **temperature probe** connected to an **interface** that is connected to a **computer** with **Logger Pro**. Looking at the materials and lab we will be using, what are the safety precautions we should take to protect ourselves and materials during the investigation?

1) In the quart-sized Ziplock bag, place a cup of milk and two tablespoons of Nesquik. Carefully take as much air out of the bag as possible, seal it shut, and mix the milk and Nesquik.
2) Fill the gallon-size Ziplock bag half full with ice. Place the temperature probe inside the bag of ice to measure the temperature of the ice. Once the temperature stabilizes, write it in Data Table 1 and take the temperature probe out of the bag.
3) Add five tablespoons of salt into the gallon-size Ziplock bag of ice.
4) Place the quart-sized bag into the gallon-sized bag. Remove most of the air in the gallon-sized bag, seal it, put on your gloves, and carefully shake and mix. Be careful not to break the bag on the inside, or you will have salty ice cream that will taste nasty. Make sure you keep shaking; otherwise, your ice cream will not be creamy and will freeze into a block of ice (not fun to eat).
5) Once the milk thickens up and turns slushy (it takes 3-5 minutes), open the gallon Ziplock bag and measure the temperature of the ice and salt mixture. Once temperature stabilizes, write this in Data Table 1.
6) Take the quart-sized Ziplock bag out, open your bag, and eat your ice cream with your spoon.

Data Table 1

	Ice by itself	Ice and salt mixture	Temperature Difference
Temperature (C°)	C°	C°	C°

Questions:

1) Why do you think the ice temperature in the gallon-sized bag dropped?

2) Why did the milk mixture freeze to ice cream?

3) From what you observed in this investigation, which water do you think will have a lower freezing point, ocean water or fresh water in a lake? Explain why.

4) What happened to the temperature of the milk mixture during the investigation? How do you know?

5) How do you think changing the size of the ice cubes or the granules of salt would change the results of the investigation?

The Solubility of Gas in a Liquid

Directions and Questions:

You will need **safety goggles**, two **glass bottles of soda**, two **balloons** (that you fit on the tops of the bottles immediately after they are opened), and a **hotplate. Looking at the materials and lab we will be using, what are the safety precautions we should take to protect ourselves and materials during the investigation?**

1) Open both bottles and fix and seal a balloon over the opening of each bottle. Place one bottle on the hotplate. Observe both bottles; what do you see happening to the balloons? Why do you think this is happening?

2) Make sure to turn the hotplate off when the experiment is over so it does not boil over. Why do you think we put lids on bottles of soda and keep them in the refrigerator?

3) Now take the other bottle, not on the hotplate, and shake it. What do you see happen to the balloon now?

4) Why is this happening?

5) What can you say about how gasses stay dissolved in a liquid solution? (**hint:** 3 things)

Using Beer's Law

Directions:

You will need **safety goggles**, an **apron**, 5 **test tubes** (label the 1, 2, 3, 4 & 5), a **test tube rack**, a **stirring rod**, two **measuring pipettes**, **.4M $NiSO_4$ solution** (try not to touch this with your skin), **distilled water**, **lens paper**, and a **Colorimeter** attached to an **interface** connected to a **computer** with **Logger Pro**. Looking at the materials and lab we will be using, what are the safety precautions we should take to protect ourselves and materials during this investigation?

1) Prepare test tube 1 with 2 mL of nickel sulfate and 8 mL of distilled water (do not mix the pipets, one for the nickel sulfate and the other for the distilled water).
2) Prepare test tube 2 with 4 mL nickel sulfate and 6 mL of distilled water.
3) In keeping with this pattern, prepare test tubes 3 and 4 with 6 mL and 8 mL of nickel sulfate and 4mL and 2 mL of distilled water respectfully. These are what we will use to determine the standard. The concentrations are already calculated for you in Data Table 1. Your teacher will make the unknown for you to determine its concentration for test tube 5.
4) You will need to open the folder Chemistry with Vernier and file #11 Beer's Law.
5) You will now need to calibrate the colorimeter by preparing a blank by filling the cuvette ¾ full with distilled water. Ensure the cuvette is wiped clean of fingerprints, moisture on the outside, and no bubbles inside. Open the lid and make sure the marks line up when placing it into the colorimeter. Close the lid and press the <or> button to select a wavelength of 635 nm (red) for this experiment. Press the CAL button until the red LED begins to flash. Then release the CAL button. When the LED stops flashing, the calibration is complete. Click "Collect." Empty the water from the cuvette.
6) Fill the cuvette ¾ full with test tube one solution. Wipe the outside, making sure there are no fingerprints or water on the outside. Place it into the colorimeter properly lined up and close the lid. Click "Keep" and type .080 in the edit box, and press enter. You should see that data point plotted on the graph. Open the lid, take out the cuvette, and discard its contents as directed by your teacher. Carefully rinse the cuvette with distilled water and get ready to repeat this procedure.

7) Repeat the procedure in #6 for test tubes 2, 3, and 4. When you click "Keep" for each of those tubes, write in the concentration in Data Table 1 for each test tube: two is .16, three is .24, and four is .32. When you have finished all four test tubes, click "Stop."
8) For Data Table 1, record the absorbance data paired with the displayed data table's concentration in Logger Pro.
9) Click the Linear Fit button to find the line that best fits your data in Logger Pro.
10) Now obtain the unknown concentration and repeat the procedure in #6 again, except you do not need to click "Collect "or "Keep" since this data will be live. When the displayed absorbance value stabilizes, record it in Data table 1. Use your graph in Logger Pro to estimate the concentration of the solution. You can move your cursor over the graph to find the absorbance value; next to it will be the floating box's concentration value. If this does not happen automatically, choose Interpolate from the Analyze menu. Write this also in Data Table 1.

Data Table 1

Test Tube	Concentration (mol/L)	Absorbance
1	.08	
2	.16	
3	.24	
4	.32	
5 (unknown)	mol/L	

Questions:

1) What do you think Beer's Law states?

2) How could we have obtained errors in the data?

3) Why did we wipe the cuvette each time we used it?

Elements Compounds and Mixtures Research

Directions and Questions:

Using your teacher's instructions, use the **internet** and your **textbook** to research elements, compounds, and mixtures, then answer the following questions.

1) What is an element, and how is it related to compounds and mixtures?

2) What are examples of elements?

3) What is a compound, and how is it related to elements and mixtures?

4) What are examples of compounds?

5) What is a mixture, and how is it related to elements and compounds?

6) What are the different categories of mixtures, and how do you tell them apart?

7) How were the different elements made?

8) So can an element be separated?

   a. Is it easy or hard? Explain.

9) How can a compound be separated?

   a. Is it easy or hard? Explain.

10) How can a mixture be separated?

    a. Is it easy or hard? Explain.

11) What does this investigation tell us about ourselves?

12) Fill in the Ven diagram below comparing and contrasting elements, compounds, and mixtures.

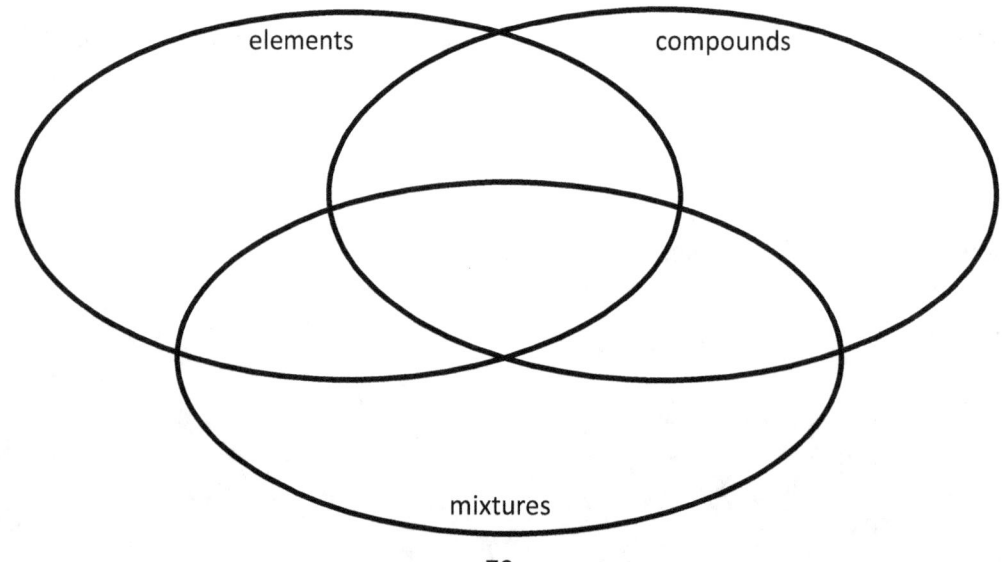

Simple Chemistry Investigations   Seven Sides Publishing

Elements Compounds and Mixtures

Directions:

You will need **aluminum foil**, **milk**, a **laser pointer**, **water**, **granite countertop** samples, **Cool Aid**, **salt** (sodium chloride), a **pencil**, **chalk** (calcium carbonate), and **muddy water**. Looking at the materials and lab we will be using, what are the safety precautions we should take to protect ourselves and materials during the investigation?

1) **Pure substances** can be **elements** or **compounds**.
    a. **Elements** are pure substances with only one type of atom found on the periodic table.
    b. **Compounds** are pure substances with two or more elements in a fixed ratio.
2) **Mixtures** have varying ratios of different substances.
    a. **Homogeneous mixtures** appear the same throughout. They are also called **solutions**. Lasers can go through **solutions**.
    b. **Heterogeneous mixtures** are different throughout. Lasers are deflected in heterogeneous mixtures.
        i. **Colloids** look like homogeneous mixtures but have large particles that do not settle out.
        ii. **Suspensions** have larger particles that do settle out.
3) Analyze each substance in Data Table 1 and classify it as either a pure substance or a mixture. Then classify the type of pure substance or mixture. This data will be filled in Data Table 1.

Data Table 1

Object	Identity (Pure or Mix)	Classification
Aluminum foil		
Milk		
Water		
Salt		
Granite countertop		
Cool aid		
Salt		
Pencil		
Chalk		
Muddy water		

Questions:

1) If you know the name of a substance, how can you tell whether it is an element or not?

2) How did you find out how milk was classified?

3) How would you classify a human? Explain why.

4) How would you classify fog?

5) How would you classify a clean atmosphere?

6) How would you classify smog?

7) How would you classify brass?

Separating Mixtures

Directions and Questions:

You will need a **magnet** in a **plastic baggy** (keep the magnet in the plastic baggy the whole time), a **wire strainer**, a **coffee filter**, a **hotplate**, **water**, **sand**, **sugar**, **marbles**, **iron filings**, **granola**, and two **beakers. Looking at the materials and lab we will be using, what are the safety precautions we should take to protect ourselves and materials during the investigation?**

1) Take the mixture of sand, sugar, marbles, iron filings, and granola and use the materials to separate the mixture into its pieces. How will you divide the marbles from the mixture?

2) How will you separate the iron filings from the mixture?

3) How will you separate the granola from the mixture?

4) How will you separate the sand from the mixture?

5) How will you separate the sugar from the mixture?

Separating Pigments

Directions:

You will need **goggles**, **scissors**, different **pens** and **markers**, an **eyedropper**, **nail polish remover** or **alcohol**, **filter paper** or **chromatography paper**, **test tubes**, a **test tube rack**, **paper clips**, and **rubber stoppers** with holes in them that fit the test tubes. **Looking at the materials and lab we will be using, what are the safety precautions we should take to protect ourselves and materials during the investigation?**

1) You are going to do chromatography today. Take the paper and cut it into strips if you have to, with a point at one end.
2) Make a dot with the pen or marker about an inch from the pointy tip.
3) Take a paper clip and bend it so there is a hook on one end that you poke through the flat end of the strip of paper and put the other end through the hole in the stopper.
4) Take the eyedropper and carefully squirt a little nail polish remover/alcohol into the bottom of a test tube. Then lower the paper with the dot pointy end down into the nail polish remover, but do not let the dot touch it. Fix the stopper at the top of the test tube. Bend the part of the paper clip that is above the stopper so that the paper does not drop any lower.
5) Repeat the process you did for #s 1-4 for all the different pens and markers you have chosen.
6) Watch the patterns of pigments that separate. The ones that move the fastest go to the top, the ones that move the slowest will be at the bottom.
7) After your pigments have separated, take them out of the test tubes and let them dry out (otherwise, all the stains may go back together again at the top of the paper).
8) Draw pictures of the color bands that came out from your pens and markers on your paper strips below:

Questions:

1) Did you see the same colors make different patterns?

2) Can the same color of ink be made of different substances?

    a. How did you see this today?

3) How could you use this information to determine which pen wrote a note if you need to?

50 + 50 Does Not Equal 100

Directions and Questions:

You will need **safety goggles**, two **400 mL beakers**, two **100 mL graduated cylinders**, sand, **marbles**, 50 mL of **water**, and 50 mL of **rubbing alcohol**. **Looking at the materials and lab we will be using, what are the safety precautions we should take to protect ourselves and materials during the investigation?**

1) Carefully measure 50 mL of water in one graduated cylinder and 50 mL of alcohol in another.
2) Carefully pour the contents of one of the graduated cylinders into the other. What is the total volume now?

3) How do you think this could have happened?

4) Fill a beaker 200 mL full with marbles and another 200 mL full with sand. Gently pour the sand into the marbles, slowly shaking the beaker of the marbles. How many milliliters of marbles and sand is there now in the beaker?

5) Where did the sand go to lose the volume?

6) How can you explain the results for #2 now?

Percent Sugar in Bubble Gum

Directions:

You will need a **scale** and **bubble gum**. **Looking at the materials and lab we will be using, what are the safety precautions we should take to protect ourselves and materials during the investigation?**

1) Find the mass of one piece of gum sitting on the wrapper; write this in Data Table 1.
2) Take the gum and chew it. As you are chewing, find the mass of the wrapper and write it in Data Table 1.
3) Subtract the wrapper's mass from the gum and wrapper's mass to get just the unchewed gum's mass; write this in Data Table 1.
4) When you notice no more flavor in the gum, put the chewed gum back on the wrapper and find the chewed gum's mass (do not forget to subtract the wrapper's mass). Write the mass of the chewed gum in Data Table 1.
5) To find the percent of the gum that was not sugar, take the chewed gum's mass and divide it by the unchewed gum's mass times 100.

Data Table 1

Mass of unchewed gum and wrapper	Mass of wrapper	Mass of unchewed gum	Mass of chewed Gum	% of gum, not sugar	% of gum that is sugar
g	g	g	g	g	g

Questions:

1) What was the percentage of gum that is sugar?

2) Where did the sugar and flavor go?

3) So when you are chewing gum, are you technically eating? Explain why.

Extension: try this for different brands of gum or even sugarless gum. Show your results below.

Virtual Investigations that go with Pure Substances and Mixtures

ExploreLearning.com:

    Colligative Properties Gizmo

    Solubility and Temperature Gizmo

    Freezing Point of Salt Water Gizmo

    Equilibrium and Pressure Gizmo

    Osmosis Gizmo

Phet.colorado.edu:

    Beer's Law

    Concentration

    Molarity

    Salts and Solubility

    Sugar and Salt Solutions

Physicsclassroom.com/Concept-Builders/Chemistry:

    Classification of Matter

Unit 5: Atomic Structure and Nuclear

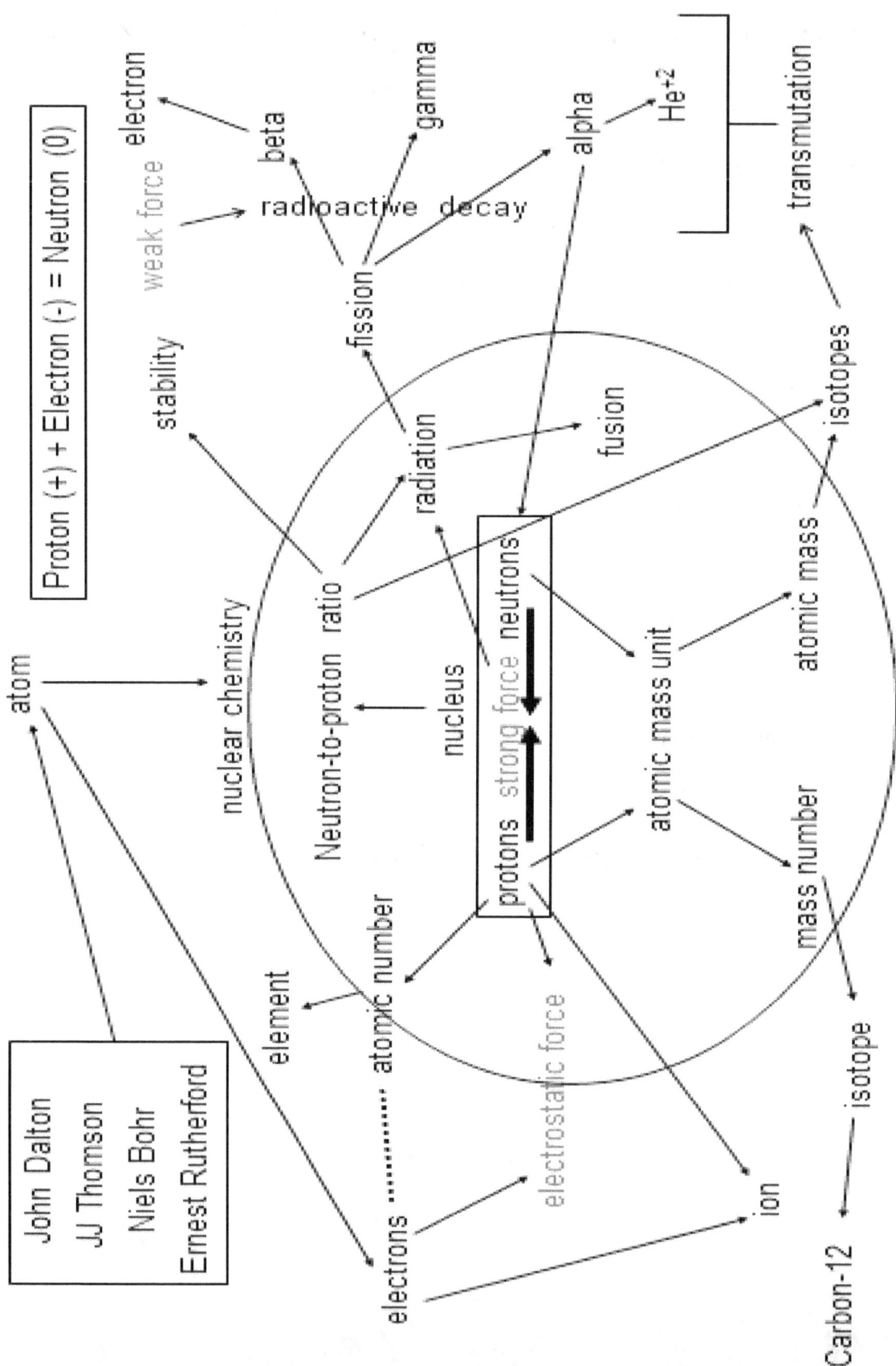

Simple History of the Atom

Directions and Questions:

Use the **internet** and your **textbook** to research the history of the development of our knowledge of the atom. Include the discoveries listed below and when they were made.

1) What are Dalton's Postulates?

2) What did JJ Thomson discover, and how did he discover it?

3) What was Ernest Rutherford's gold foil experiment, and what did it show?

4) What was Bohr's model of the atom?

Picture of Actual Atoms

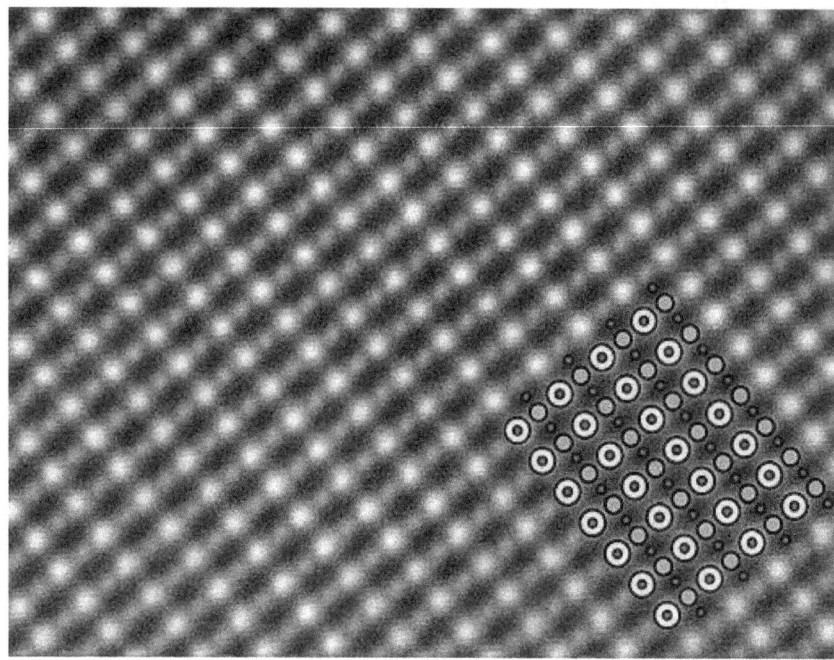

Atom resolution STEM image by Magnunor (Own work) CC BY-SA 4.0.

Directions and Questions:

Use the **internet** to research the history of the development of modern atomic theory. Include models of Dalton's Postulates, Thomson's discovery of the electron properties, Rutherford's nuclear atom, Bohr's nuclear atom, and Heisenberg's Uncertainty Principle to show how science developed the atomic theory we have today. Use whatever digital media your teacher chooses to show this, then answer the questions that follow in your project.

1) The pixels of the universe are illusions. Atoms, the pixels of our universe, appear as solid balls, but are they?

2) Do you think subatomic particles experience time?

3) Is there anything solid inside the atom?

4) If atoms are mostly empty space, why do they look solid?

5) Why do we consider the atomic theory a theory? Why is it not a law or hypothesis?

Scale Model of a Hydrogen Atom

Directions and Questions:

You will need a **golf ball**, a **bead**, and a **large field** or **parking lot. Looking at the materials and lab we will be using, what are the safety precautions we should take to protect ourselves and materials during this investigation?**

1) Walk out to a large field or parking lot, at least the size of a football field. Keep in mind the space you use still may be too small to be a scale model. You will make a model of a hydrogen atom with 1 proton and 1 electron.
2) On one edge, take a small red bead representing an electron and put it somewhere where you can see it (hang it on a fence or a tiny branch).
3) Walk at least 100 yards away; if you have more room, you can use that. Hold up the golf ball, which is a proton, and answer the questions that follow. Can you see the bead?

4) This distance is how far away the closest electron speeds around the proton. The speed approaches the speed of light. It moves so fast that it makes a ball the size of a football stadium. If you have ever seen a fan moving fast, does it look like a disk? But is it a disk?

    a. So we must ask ourselves, the atom looks like a ball, but is it a ball? Explain why.

5) The illusion of the electron is a similar type of illusion. When an object moves close to the speed of light, time stops for that object. Quantum physics allows things without time to be everywhere they could typically be at the same time. So, why is the electron said to be everywhere in the electron cloud at the same time?

6) What do you see between the proton and electron?

7) What would happen to the electron if time were to stop?

8) If that would happen to the electron, what would happen to the atom?

9) The proton is structured similarly to this. Quarks spin and orbit inside the proton close to the speed of light, just like the electron orbits around the atom. Knowing this, how does the atom show $E=mc^2$?

10) What would happen to all atoms in the area where time stops?

11) This answer is why we call this the space-time continuum. You cannot have space without time; this is how black holes form. When time disappears, so does space. What color is a black hole? Explain why.

12) Is there any space at the bottom of the black hole?

13) What do you think the temperature is at the bottom of the black hole?

    a. What is the volume of gas at this temperature?

14) This model also shows that the pixel of our universe is an illusion. If the pixel of our universe is an illusion, what does that say about our universe?

15) What do you think is the ultimate reality?

16) With our laws of physics, are we allowed to know what ultimate reality is?

Simple Chemistry Investigations   Seven Sides Publishing

Model of an Atom Showing the Illusion

Directions and Questions:

Take a **golf ball** and **small bead** out to a **large parking lot** or **field** larger than a football/soccer field. Place the small bead (electron) on the edge of the field/parking lot. Walk to the center of the field/lot with the golf ball (proton). The electron spins around the proton this far away, so fast that it makes it look like a ball about the size of a football Stadium.

1) What is between the electron and proton?

2) What happens to the atom if the electron stops moving?

3) What happens to the electron if time stops?

4) What happens to the atom if time stops?

The proton is in the center of the atom (nucleus) and is made up of quarks that spin around really fast like the electron does around the atom giving the illusion that the proton is whole. These quarks are ultimately made of waves of energy.

The electron has photons, which are both particles and waves.

5) What happens if you stop the electron motion?

6) What happens to the part of the universe where you stop time?

This answer is why time and space are connected in space-time. You cannot have one without the other.

When you approach the speed of light, time stops. Black holes accelerate matter into the center of it, and as this happens, the object either leaves the universe or is changed into energy. Time and space do not exist in the black hole's center, which is why it is black. All our laws of physics break down inside the black hole.

7) What could happen if you pass into a black hole?

8) Do you think electrons experience time?

9) Do you think time exists inside of the atom?

E=mc² says matter can be converted into energy, and energy can be converted into matter. The way the atom is structured shows us this is not only possible but also real.

String Theory shows that each particle in the atom is composed of a vibrating string. Each particle has its own vibration signature that defines that particle. String Theory also says the other seven dimensions of this universe that we cannot perceive are inside the atom.

10) What do you think could be going on in there?

11) The whole universe has and can fit inside of what structure?

    a. We know all this is possible because the atom is mostly made up of what?

Simple Chemistry Investigations Seven Sides Publishing

Building Bohr Models

Directions:

You will need a **film case** filled with **three colors of beads** (I use red, white, and blue) and a **periodic table**. **Looking at the materials and lab we will be using, what are the safety precautions we should take to protect ourselves and materials during the investigation?**

1) Designate which color bead represents which part of the atom. Here is an example: Red beads are electrons, white beads are neutrons, and blue beads are protons. I use these colors because this is how my periodic table was colored. These colors help my students learn how and why the periodic table works.
2) Carefully empty the film case. As the students empty those on the table, some beads will bounce out. I call this radiation because radiation is particles and energy coming out of an atom.
3) Also know that a neutron is the combination of a proton and an electron. This is why it is neutral. Proton (p+) + Electron (e-) = Neutron (n°) or 1+ + 1- = 0.
4) Place the empty film canister in the center circle labeled the nucleus on page 90.
5) Now make a **Hydrogen atom** by looking at the periodic table and having the student look at the atomic number; this tells us how many protons are in that atom. What defines the element is the number of protons. The atomic number for Hydrogen is 1. So place one blue proton in the nucleus (film case). The number of protons also tells you how many electrons there will be in a neutral atom. So place one red bead in the first orbit closest to the nucleus. Protons and neutrons have mass, and electrons do not. Look at the average atomic mass at the bottom of the element's box. Round it to the closest whole number; this tells you the mass of most of the atoms of that element. Since the mass is one and we have one bead in the nucleus, we have completed the hydrogen 1 (H-1) isotope.
6) Add a neutron (white bead) to the nucleus; this makes a Hydrogen 2 (H-2) isotope. One proton and one neutron in the nucleus (2 beads) give us a mass of 2. Hydrogen 2 isotope is another version of the same element.
7) Next, make a **Helium atom**. Look at the periodic table and find Helium. How many protons does it have (look at the atomic number)? Add a blue bead to the nucleus to give it 2 protons.

8) How many electrons does Helium need to have? Add this to the first circle but on the opposite side from the other electron.

9) Why would the electron be there?

10) Now, look at the mass of the Helium. How many does it have (remember to round to the whole number)?

11) How many beads are in your nucleus?

12) Add two white beads in the nucleus (film canister) because the average mass is 4; this makes a Helium 4 (He-4) isotope.

13) Now take one neutron (white bead) out to make Helium 3 (He-3) isotope; this is what is on the moon. Scientists will want this to cause fusion reactions to produce electricity in the future.

14) We have now filled the first energy level of electrons. Notice we are on the right end of the periodic table. No more electrons can go into this circle.

15) We will now make a **Lithium atom**. How many protons does it have? Place a blue bead in the nucleus (film canister) until you match that number of protons.

16) How many electrons does Lithium have? Place an electron (red bead) on the second circle. We are now at the second energy level.

17) Look at the average mass of Lithium. Add neutrons (white beads) to the nucleus until you have matched the mass of Lithium.

18) Now go and make the other isotopes, **Carbon 12** (which is in all life) and **Carbon 14** (which is radioactive and starts to decay when an organism dies), which are essential to life. **Carbon 14** turns into **Nitrogen** by taking an electron out of a neutron (a proton and an electron); this leaves a proton in the nucleus, making it a Nitrogen atom.

19) Then build **Fluorine**, **Neon**. Neon fills the second energy level, so no more electrons can fit here to make the next elements. Where do you think they will go?

20) Now build **Sodium**, **Magnesium**, and **Potassium**.

Bohr Model

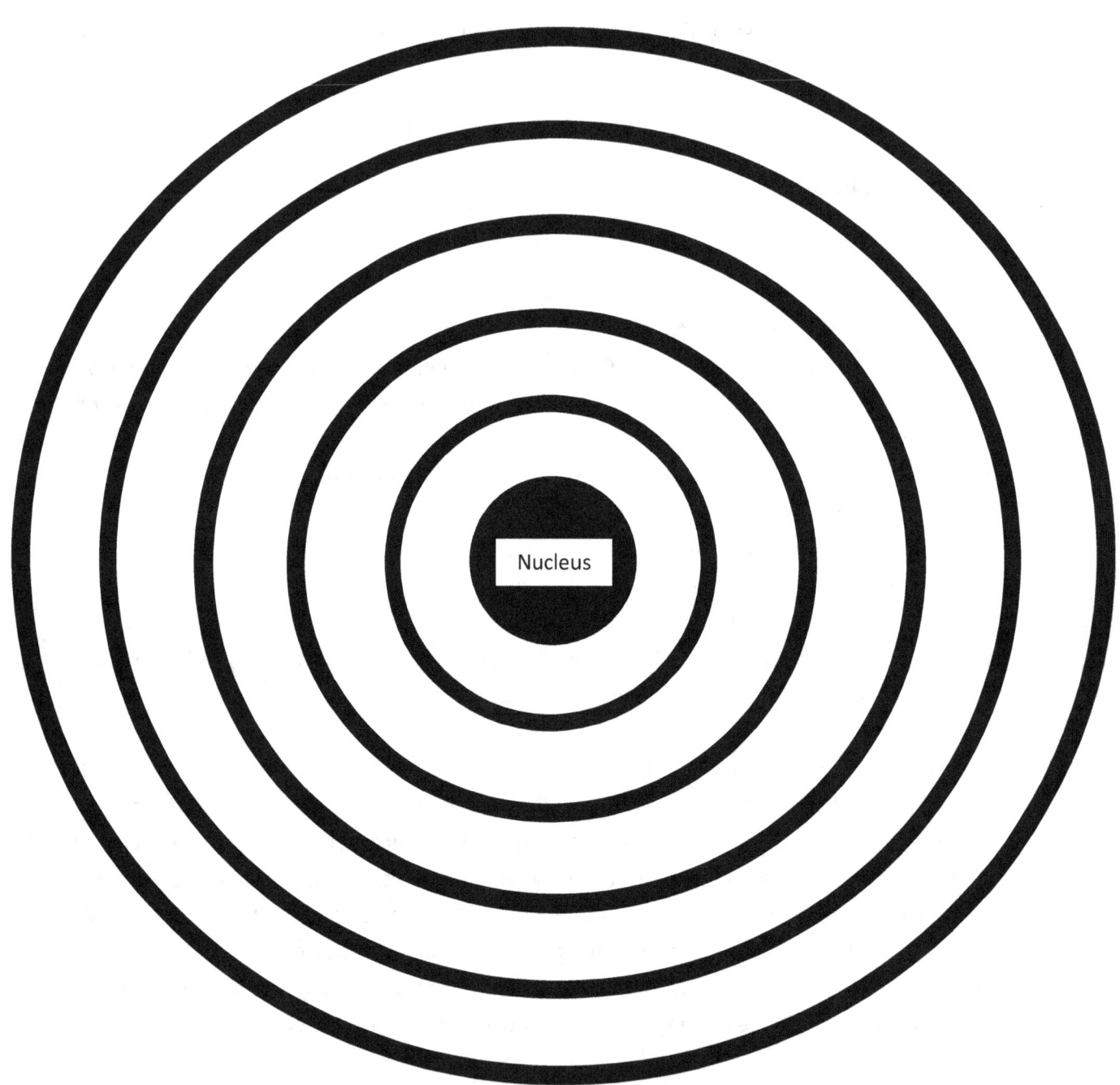

Questions:

1) How is the atomic mass determined?

2) How do you find out the number of protons for an element?

3) How do you find out the number of neutrons for an element?

4) How do you find out the number of electrons for each element?

5) How many electrons can go into the first energy level?

6) How many electrons can go into the second energy level?

7) How is the periodic table structured to tell us about the atoms of each element?

8) How does this model not accurately show where electrons orbit in atoms?

Half-life of Pennies

Directions:

You will need a **plastic tub** (about the size of a shoebox) with a **lid** and 100 **pennies. Looking at the materials and lab we will be using, what are the safety precautions we should take to protect ourselves and materials during the investigation?**

1) The half-life of a radioactive element is how long it takes for half the atoms to change into another element that is stable as they go through radioactive decay. Since pennies have two sides when flipped, almost half will land on heads; and nearly half will land on tails.
2) In your tub, place all 100 pennies heads up. These will represent your radioactive isotopes.
3) Place the lid on your box, shake it and count to 10.
4) Lift the lid and take out all the pennies that have landed tails up. Count them and fill in Data Table 1. Subtract the number of tails from 100 to show the number still heads up.
5) Now only the pennies that are heads up are still in the tub. Repeat the procedure in #s 3-4 six more times unless all your pennies turned up tails before that.
6) Graph your data from Data Table 1 on Graph 1
7) Your teacher will give each group a number and write your data into Data Table 2 for your group number. Get the other data for the other groups and write them in Data Table 2.
8) Average each half-life for all the groups by adding up their data and dividing by the number of groups.
9) Graph the averages you have for the class in Data Table 2 on Graph 2.

Data Table 1

Shaking Time (s)	# of Heads	# of Tails taken out
10		
20		
30		
40		
50		
60		
70		

Data Table 2

Groups	# of Heads at 0 (s)	# of Heads at 10 (s)	# of Heads at 20 (s)	# of Heads at 30 (s)	# of Heads at 40 (s)	# of Heads at 50 (s)	# of Heads at 60 (s)	# of Heads at 70 (s)
1	100							
2	100							
3	100							
4	100							
5	100							
6	100							
7	100							
8	100							
9	100							
10	100							
11	100							
12	100							
13	100							
14	100							
15	100							
Average	100							

Graph 1

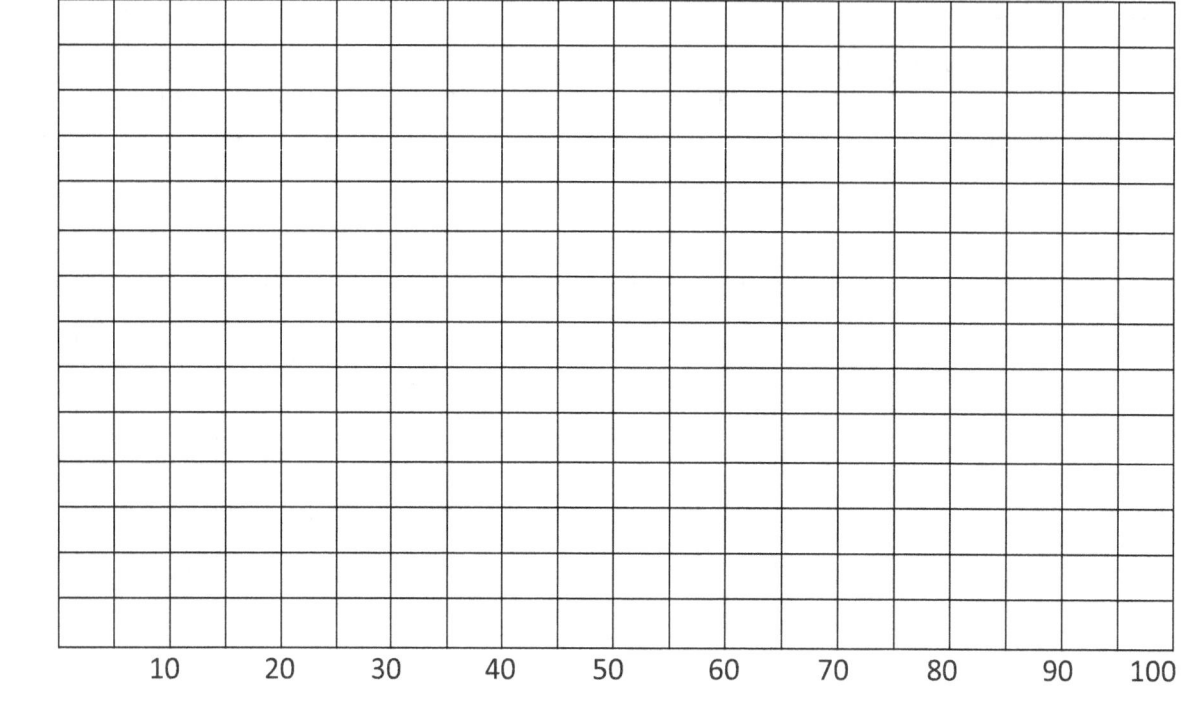

Graph 2

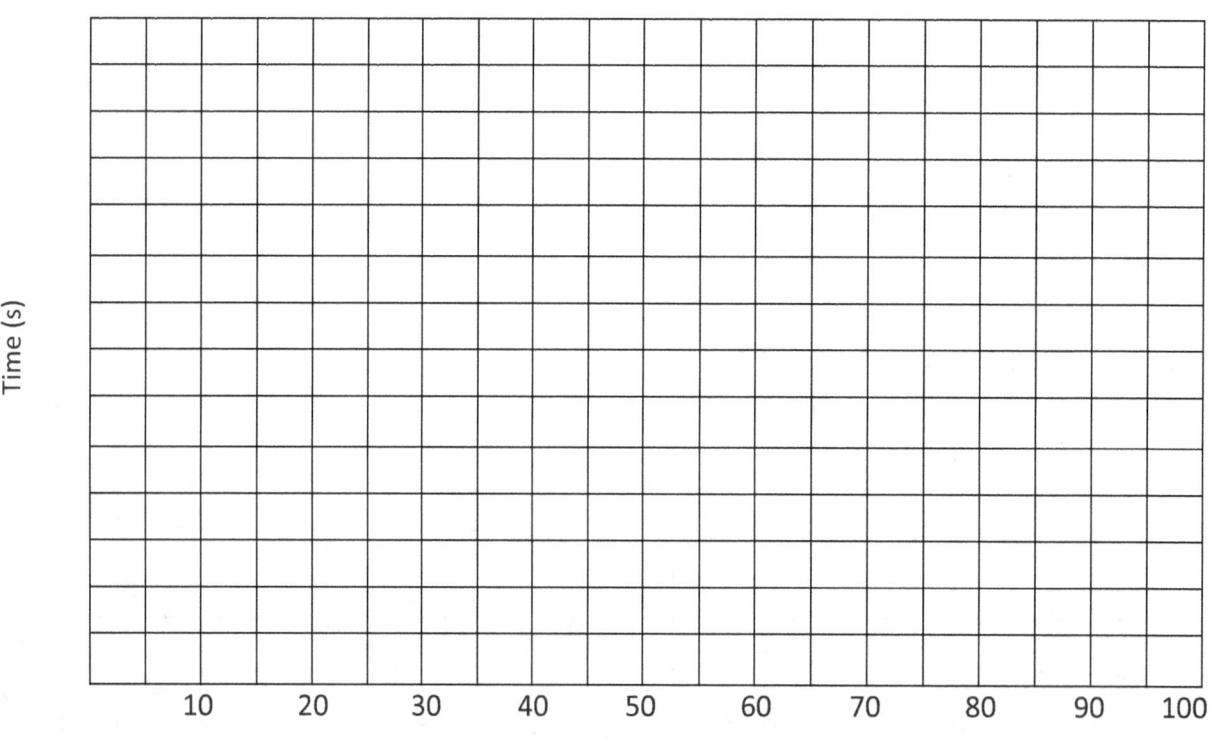

Questions:

1) What do the 10 seconds represent?

2) Which side of the coin was a stable isotope?

3) Which side of the coin was the unstable isotope?

4) What represented the radioactive decay?

5) How much of the atoms decay during each half-life?

6) How many half-lives until you should run out of pennies if all things go as expected?

7) If you start with 100 pennies and a half-life goes by every 10 seconds, how many pennies should there be after 40 seconds?

   a. How close was this to the results of your group?

   b. How close was this to the class results?

   c. Why should the class's results be closer than your group's to the expected numbers?

Calculating Nuclear Half-life Decay

Directions and Questions:

Use **colored pencils** to color in the graph below as you follow the directions to simulate what half-life looks like.

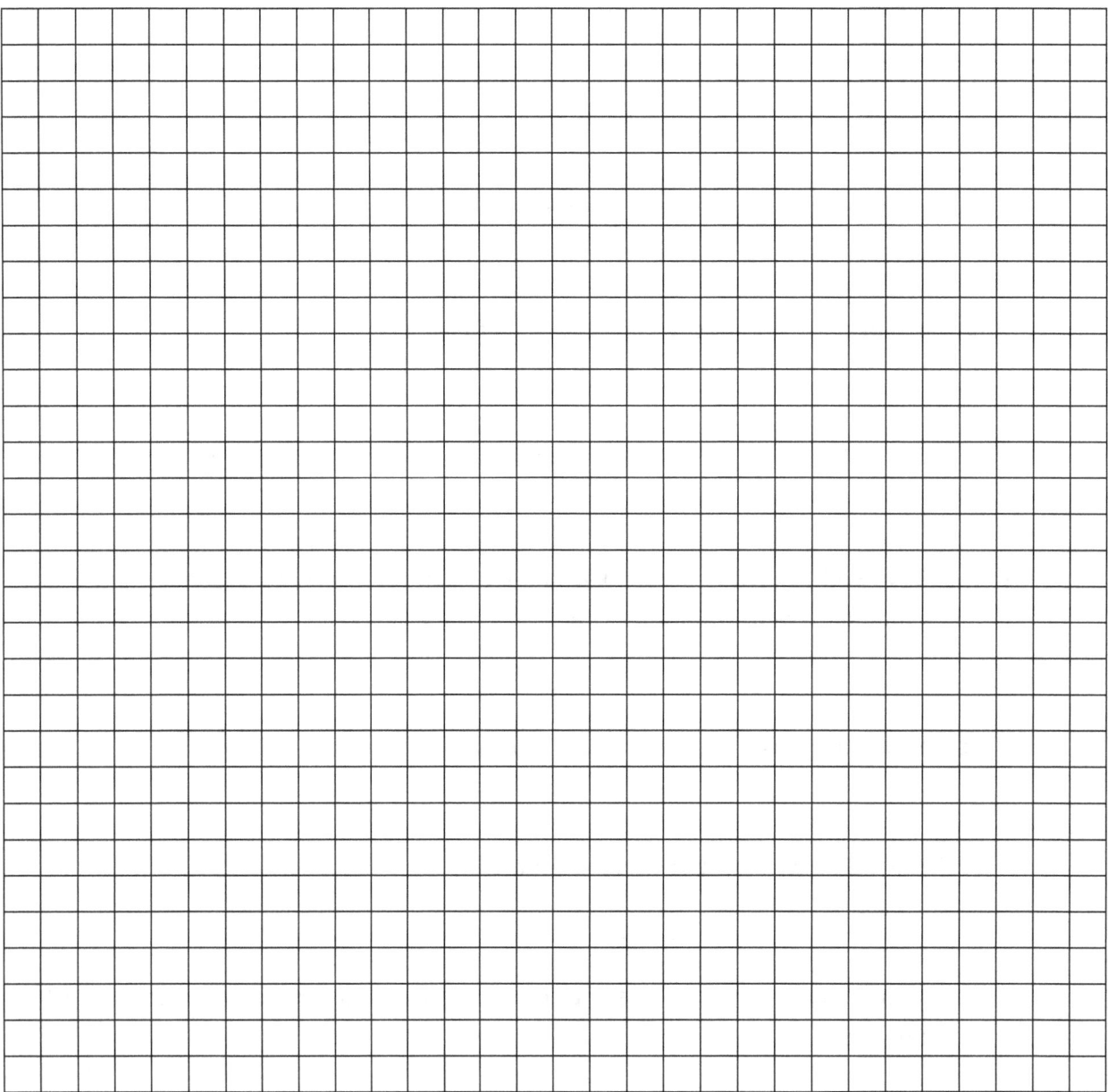

1) There are 900 squares above. Use a red colored pencil to shade in half of the squares. How many squares are left after one half-life?

2) Use a blue colored pencil to shade in half of the squares that are left; how many squares are left after two half-lives?

3) Use a green colored pencil and color in half of the squares now left. How many squares are left after three half-lives?

4) Use a yellow colored pencil and color in half of the squares now left. How many squares are left after four half-lives?

5) Use a purple colored pencil and color in half of the squares now left. How many squares are left after five half-lives?

6) Use a brown colored pencil and color in half of the squares now left. How many squares are left after six half-lives?

7) Use a black colored pencil and color in half of the squares now left. How many squares are left after seven half-lives?

8) Suppose each square represented one atom of a substance that is decomposing. How many half-lives could go by until we should expect all the atoms to be gone? Explain why.

9) If each half-life was five days, how long would it take for there to be 14 squares left?

10) If each half-life was 500 years, and there were 225 squares left, how much time went by?

Nuclear Decay

Directions:

Use the **internet** and your **textbook** to research nuclear decay. Develop a model for each type of nuclear decay (alpha, beta, gamma, and positron emission) illustrating the changes in the atom's nucleus and energy and particles released during radioactive decay. Describe how each decay happens and give examples of when they happen below.

Alpha Decay

Beta Decay

Gamma Decay

Positron Emission

Nuclear Isotopes

Directions and Questions:

Use the **internet** and your **textbook** to research how nuclear isotopes give off nuclear radiation, describe the characteristics of nuclear radiation, and show how we use it in different ways.

1) Why do some element's isotopes give off nuclear radiation?

2) What are alpha particles, and where do they come from (make sure to show a balanced reaction)?

3) What are beta particles, and where do they come from (make sure to show a balanced reaction)?

4) What are gamma rays, and where do they come from (make sure to show a balanced reaction)?

5) Which of these three is the most dangerous to life? Explain why.

6) What is radioactive dating, and how accurate is it?

7) Which radioactive isotope do we use to determine the age of a dead organism?

   a. What is its half-life, and how long can we use it after the organism dies?

8) Which radioactive isotopes help us find the age of older fossils?

9) How old is the oldest rock scientists have found?

10) What does that tell us about the age of the Earth?

11) How do we use nuclear phenomena in medicine?

   a. Would these work if the atoms were solid, or can they only happen because the atoms are mostly empty space?

12) How do we use nuclear phenomena to make electricity?

Nuclear Fission and Fusion

Directions and Questions:

Use the **internet** and your **textbook** to research and explain how nuclear fission and fusion happen.

1) Draw a diagram of nuclear fission.

2) What are the element isotopes involved?

3) How does the reaction get started (what goes into the reaction)?

4) What comes out of the fission reaction?

5) How does this lead to a chain reaction?

6) How do we slow this reaction down?

7) Draw a diagram of nuclear fusion.

8) What are the element isotopes involved?

9) How is fusion different than fission?

10) What goes into the fusion reaction?

11) What comes out of the fusion reaction?

12) Which one (fission or fusion) has a waste product that is harmful to the environment?

13) Which one (fission or fusion) does not have any harmful waste products?

14) Which one (fission or fusion) do we know how to use to make usable energy?

15) Which one (fission or fusion) is being researched to make clean energy but not usable yet?

Simple Chemistry Investigations   Seven Sides Publishing

Nuclear Chain Reactions

Directions:

You will need a large number of **dominos** and a **stopwatch. Looking at the materials and lab we will be using, what are the safety precautions we should take to protect ourselves and materials during the investigation?**

1) Take half the dominos and set them up in one line so that when you knock the first one over, the first domino will hit the second, the second will hit the third, and so on until they all fall; this is Domino Set 1. Set 1 is like a nuclear chain reaction in a nuclear reactor.
2) Take the other set of dominos and set them up so that the first domino will knock down two dominos, and those two dominos will knock down 4, and those four will knock down 8, and so on until they are all used up; this is Domino Set 2. Set 2 is like a nuclear bomb blowing up.
3) Take your stopwatch and time how long it takes to knock down the dominos in set 1. How long did it take to knock down domino set 1?

4) Then find out how long it takes to knock down the dominos in set 2. How long did it take to knock down domino set 2?

Questions:

1) Which reaction took longer?

2) Why do you think we want nuclear chain reactions to go slower in a nuclear reactor that produces electricity?

3) What would happen if the nuclear reactor acted more like Domino Set 2?

4) How do you think we slow down this reaction and cool it off so it does not blow up?

Simple Chemistry Investigations — Seven Sides Publishing

Nuclear Reactor

Directions and Questions:

Use the **internet** and your **textbook** to research how nuclear reactors work.

1) Draw a diagram of the main parts of a nuclear reactor showing how it works.

2) Where does the reaction take place in the reactor?

3) How is the water there then used?

4) How does this make electricity?

5) How do we cool off the reactor so it does not explode?

6) How do we slow the reaction down so it does not explode?

7) What is the fallout from a nuclear reactor meltdown?

8) What do we do with the waste products?

9) Why does the US prohibit the production of any new nuclear power plants?

10) How do other countries dispose of their nuclear waste?

11) What are the positive and negative effects of using nuclear power plants to produce electricity?

12) Why do we not want developing countries to develop this technology?

Smoke Detector

Directions and Questions:

Use the **internet** and your **textbook** to research and explain how a smoke detector works with nuclear radiation.

1) Draw a diagram of how the radioactive isotopes work with the gases in the air to complete a circuit in the smoke detector.

2) Which elements are involved?

3) How are the elements interacting?

4) What happens when smoke gets in the way?

Applications of Nuclear Phenomena

Directions and Questions:

Use the **internet** and your **textbook** to research solar cells, different types of medical imaging, and radiation therapy. Use this information to answer the questions below.

1) Develop a model for a solar cell and explain how it can collect and transfer energy to make an electric current.

2) Describe three different kinds of medical imaging that use nuclear technology and explain how they can capture, transmit, and interpret information for doctors to diagnose patients.

3) Describe three different kinds of radiation therapy used to help treat disease. Explain how each nuclear technology works to focus energy on specific body areas.

Virtual Investigations that go with Atomic Structure and Nuclear

ExploreLearning.com:

    Element Builder Gizmo

    Average Atomic Mass Gizmo

    Isotopes Gizmo

    Nuclear Decay Gizmo

    Half-Life Gizmo

    Nuclear Reactions Gizmo

Phet.colorado.edu:

    Build an Atom

    Isotopes and Atomic Mass

    Models of Hydrogen Atom

    Alpha Decay

    Beta Decay

    Nuclear Fission

    Radioactive Dating Game

    Rutherford Scattering

    Stren-Gerlach Experiment

Physicsclassroom.com/Concept-Builder/Chemistry:

    Name that Element

    Isotopes

    Nuclear Decay

Unit 6: Electrons

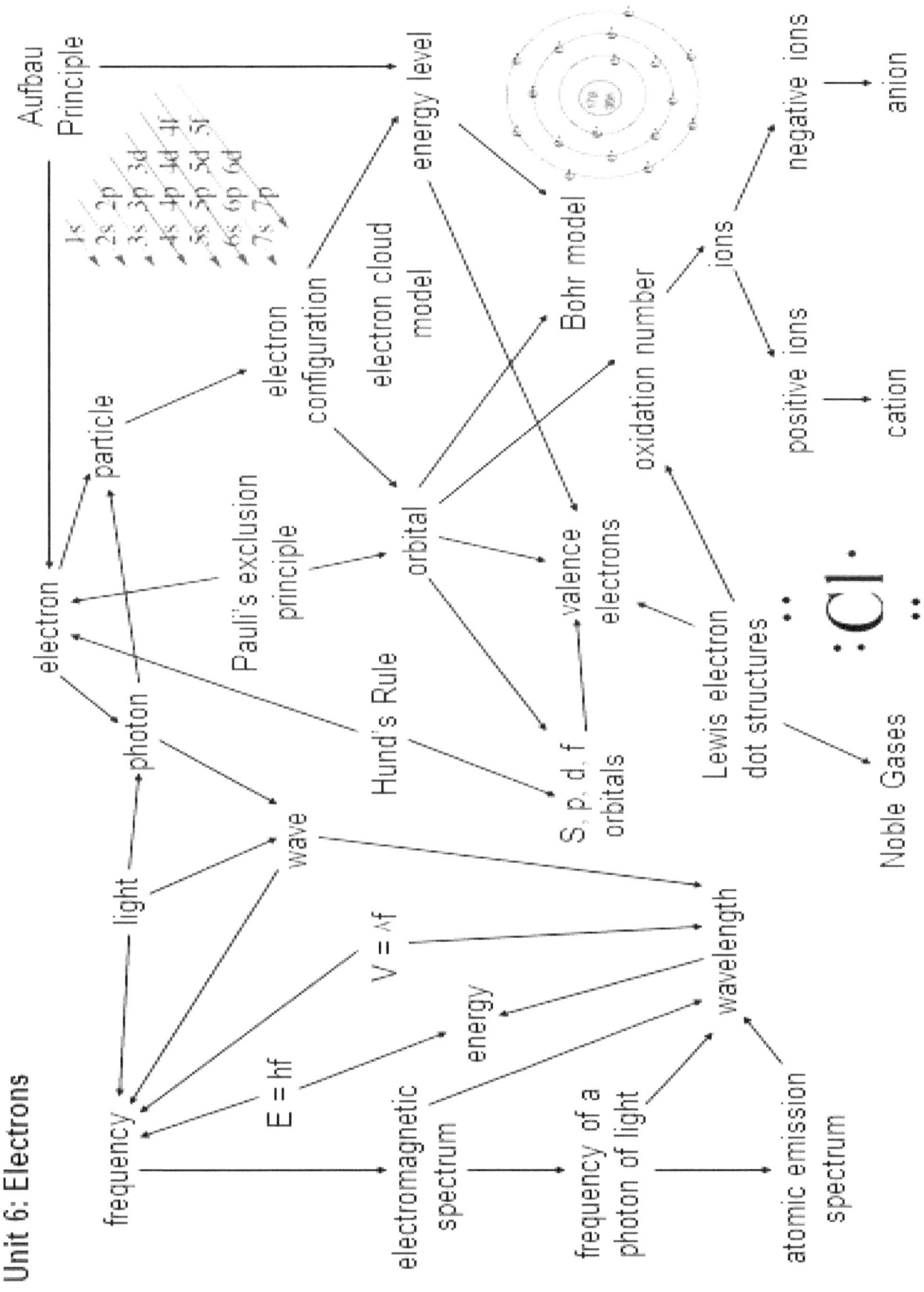

Niels Bohr's Contributions of Electrons

Directions and Questions:

Use the **internet** and your **textbook** to research the contribution of Niels Bohr to chemistry and physics. Include the information listed below.

1) What was Bohr's model of the atom? Draw a model and explain it.

2) What did Niels Bohr discover about electrons? Draw models showing how electrons can move from one energy level to another.

3) How did he explain the structure of the periodic table?

4) What is the wave-particle duality Bohr formulated in 1927?

a. What seems unreal about what he found?

5) Describe the Heisenberg's Uncertainty Principle.

a. What does it tell us about where individual electrons are in the atom?

Seats and Electron Positions (One Class)

Directions:

Students will model *electrons*, and the **seats** in your room will represent *positions in orbitals*. The rows of the seats represent the energy levels. Electrons filling orbitals act a lot like how people normally take seats on a bus.

The Rules:

1) An electron wants to sit in the closest position in an orbital that is empty.
2) If all the positions in the orbitals have a seat filled, then they will want to occupy the closest potion in the orbital with an opening.
3) When two electrons are in the same position of an orbital, they move in opposite directions. In your model, have the seats facing the opposite directions at each position. In each orbital, students will fill in the **forward-facing seats first** at each position, **then the backward-facing seats** from **closest to farthest** positions (from the door) in the orbital.

Because this is so complicated, it would be good to use the whole class to act this out and fill in seats like electrons would by following the rules and patterns.

Teachers, you will need to set up your room (however big your class is) like the electron configuration. Each row represents an energy level, and each row section represents an orbital. Remember to have one seat facing forward and one seat facing backward at each position. It will be helpful to have a **Periodic Table** to help guide you and the students. Then have the students fill in with the following pattern until you run out of students:

1) The first row has the first orbital with one position and two seats **1s**. Have the first 2 students sit there following the rules. As they each sit separately, have the whole class call out the neutral element the configuration is for. When the first student sits, have the class call out **hydrogen**. When the second student sits, have the class call out **Helium**.
2) The second row has two orbitals. The first one has one position and two seats **2s**. Have the next 2 students sit there one at a time, following the rules. As they each sit, have the class call out **Lithium** and then **Berylium**.
    a. The next orbital has three positions with six seats **2p**. Have the next 6 students sit there one by one filling the seats according to the rules. As they sit, have the class call out **Boron, Carbon, Nitrogen, Oxygen, Fluorine,** and **Neon**.

3) The third row has three orbitals. The first one has one position with two seats **3s**. Have the next 2 students sit there following the rules. Call out **Sodium** and **Magnesium** as they sit. **Keep having the class call out each element on the periodic table as each student sits until you use up all your students or do the whole periodic table.**
    a. The next orbital has three positions with six seats **3p**. Have the next 6 students sit there one by one, filling in the seats according to the rules. The rest of this row will have to wait until the first part of the fourth row is filled because it is closer to the entrance.
4) The fourth row has four orbitals. The first orbital has one position with two seats **4s**. Have the next 2 students sit there following the rules.
    a. The next part is tricky because we will now go back and fill in the third row; the third orbital here has five positions with ten seats **3d**. Have the next 10 students sit there one by one, filling in the seats according to the rules.
    b. Now we come back to the fourth row, where the second orbital has three positions with six seats **4p**. If you have enough students have the next 6 students sit there one by one filling the seats according to the rules.
5) The fifth row has four orbitals. The first has one position with two seats **5s**. Have the next 2 students sit there following the rules.
6) If you want to continue, you need to combine classes and do the next activity using a bigger room.
7) Looking at this pattern, what do you think will be the name of the next orbital? _____

Questions:

1) How does this pattern follow what you just did as a class? Draw in diagonal arrows to show the order you filled in the orbitals.

    1s

    2s   2p

    3s   3p   3d

    4s   4p   4d   4f

    5s   5p   5d   5f

    6s   6p   6d

    7s   7p

2) How far was your class able to go down the periodic table?

3) Look at the periodic table and the pattern you have been following. How is the periodic table set up to show the orbitals electrons occupy?

4) What element would your class be if the number of students you have represented a neutral atom?

5) How does this model show how the electrons orbit a nucleus?

6) How is this model inaccurate in showing electron orbital configuration?

Seats and Electron Positions (Many Classes)

Directions:

This activity is meant to be done after each class has practiced by themselves in the previous activity. Multiple classes need to come together to make one big class.

Students will model _electrons_, and the **seats** in your room will represent _positions in orbitals_. The rows of the seats represent the energy levels. Electrons filling orbitals act a lot like how people normally take seats on a bus.

The Rules:

1) An electron wants to sit in the closest position in an orbital that is empty.
2) If all the positions in the orbitals have a seat filled, then they will want to occupy the closest potion in the orbital with an opening.
3) When two electrons are in the same position of an orbital, they move in opposite directions. In your model, have the seats facing the opposite directions at each position. In each orbital, students will fill in the **forward-facing seats first** at each position, **then the backward-facing seats** from **closest to farthest** (from the door) positions in the orbital.

Because this is so complicated, it would be good to use the whole class or multiple classes (practice with your class first before putting other classes together) to act this out and fill in seats like electrons would by following the rules and patterns.

<u>Teachers</u>, you will need to set up a **large room** like a gym, cafeteria, or multipurpose area like the electron configuration. Each row represents an energy level, and each section of a row represents an orbital. Remember to have one seat facing forward and one seat facing backward at each position. It will be helpful to have a **Periodic Table** to help guide you and the students. Then have the students fill in the following pattern:

1) The first row has the first orbital with one position and two seats **1s**. Have the first 2 students sit there following the rules. As they each sit separately, have the whole class call out the neutral element the configuration is for. When the first student sits, have the class call out **hydrogen**. When the second student sits, have the class call out **Helium**.
2) The second row has two orbitals. The first one has one position with two seats **2s**. Have the next 2 students sit there one at a time, following the rules. As they each sit, have the class call out **Lithium** and then **Berylium**.

a. The next orbital has three positions with six seats **2p**. Have the next 6 students sit there one by one filling the seats according to the rules. As they sit, have the students call out **Boron, Carbon, Nitrogen, Oxygen, Fluorine,** and **Neon**.

3) The third row has three orbitals. The first one has one position with two seats **3s**. Have the next two students sit there following the rules. Call out **Sodium** and **Magnesium** as they sit. **Keep having the class call out each element on the periodic table as each student sits until you use up all your students or do the whole periodic table.**

    a. The next orbital has three positions with six seats **3p**. Have the next 6 students sit there one by one, filling in the seats according to the rules. The rest of this row will have to wait until the first part of the fourth row is filled because it is closer to the entrance.

4) The fourth row has four orbitals. The first orbital has one position with two seats **4s**. Have the next 2 students sit there following the rules.

    a. The next part is tricky because we will now go back and fill in the third row; the third orbital here has five positions with ten seats **3d**. Have the next 10 students sit there one by one, filling in the seats according to the rules.

    b. Now we come back to the fourth row, where the second orbital has three positions with six seats **4p**. If you have enough students have the next 6 students sit there one by one filling the seats according to the rules.

5) The fifth row has four orbitals. The first has one position with two seats **5s.** Have the next 2 students sit there following the rules.

    a. We are getting tricky again and going back to the fourth row to fill in the next orbital that has five positions with ten seats **4d**. Have the next ten students sit there one by one, filling in the seats according to the rules.

    b. Then we will go back to the fifth row, where the next orbital has three positions with six seats **5p**. Have the next six students sit there one by one, filling in the seats according to the rules.

6) The sixth row has four orbitals. The first one has one position with two seats **6s**. Have the next two students sit there according to the rules.

    a. Now it gets even trickier. We must go back to the 4th row to fill in the last orbital with seven positions with fourteen seats **4f**. Have the next 14 students fill in this section one by one according to the rules.

    b. Now we can move up to the fifth row and fill in the third orbital with five positions and ten seats **5d**. Have the next 10 students fill in this section one by one, filling in the seats according to the rules.

    c. Now we can fill in the second orbital of the 6th row that has three positions with six seats at **6p**. Have the next 6 students fill in this section one by one according to the rules.

7) Looking at this pattern, what do you think will be the name of the next orbital? _____ Keep following the pattern until you run out of students or use multiple classes to fill in the entire periodic table.

Questions:

1) How far were your classes able to go down the periodic table?

2) Look at the periodic table and the pattern you have been following. How is the periodic table set up to show the orbitals electrons occupy?

3) What element would your class be if the number of students you have represented a neutral atom?

4) How does this model show how the electrons orbit a nucleus?

5) How is this model inaccurate in showing electron orbital configuration?

Atomic Structure Representations

Isotope	Symbol	Bohr Model	Electron Configuration	Noble Gas Electron Configuration	Orbital Diagrams	Lewis Dot Structure
Carbon-14	n = p = e =					
	n = p = e = 34				↑↓ ↑↓ ↑↓ ↑↓ ↑↓ ↑↓ ↑↓ ↑ ↑ ↑ (1s 2s 2p 3s 3p 4s 3d 4p 5s)	
	n = p = e = 14					
	n = p = e = 31					:N̈: (with dots below)
	n = p = e =					
Sodium-23	n = p = e = 40		$1s^2 2s^2 2p^6 3s^2 3p^6 4s^2$			
	n = p = e = 35			[Ne] $3s^2 3p^5$		
	B n = p = e = 11 5					

Summary of Electrons in each Energy Level

Energy Level N = 1, 2, 3, …	Sub-Level (s, p, d, f)	Orbital (s=1, p=3, d=5, f=7)	# of Electrons 2e- / orbital
1			
2			
3			
4			
5			
6			
7			

Electron Basics

Directions:
Use what you have learned to answer these questions. If you do not know how to answer the question, use the **internet** to help you.

1) How does an electron move to a higher energy level?

2) How does an electron move to a lower energy level?

3) What is the relationship between a photon and an electron?

4) What is the duality of light?

   a. This duality is the basis for String Theory, showing all particles are vibrating strings or waves.

5) What is given off when atoms hit each other so hard that the electrons fall off?

6) How is plasma different than gas?

7) What is the electron cloud? Describe it.

8) Why do you think electrons appear everywhere in the electron cloud at the same time?

Uses of the Electromagnetic Spectrum

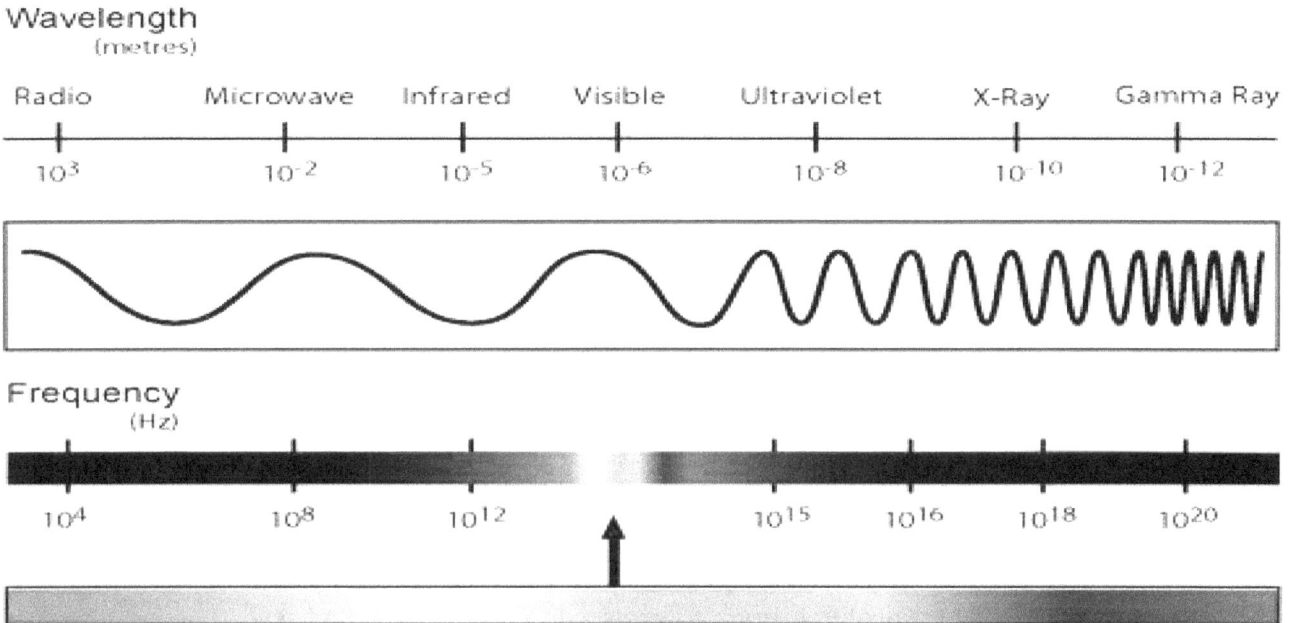

Directions and Questions:

Use the Electromagnetic Spectrum above and the **internet** to answer the questions below.

1) Electromagnetic waves have higher energy with shorter wavelengths and lower energy with longer wavelengths. Which waves above would have the highest energy, also being the most dangerous?

2) Which waves above have the lowest energy and are the most harmless?

3) Where is visible light on this spectrum?

4) Why do you think cell phones use microwaves to communicate?

5) Longer wavelengths can travel around corners easier. Which would travel farther, radio waves or microwaves?

6) Which would you see more of radio towers or cell phone towers? Explain why.

7) Why do you think remote controls use infrared waves to control TVs, drones, and remote control cars?

8) How do you think a pilot in Arizona controls a drone in the Middle East?

9) Why do you think the doctor puts a lead shield over you when you get an x-ray?

10) What do you think will give you a more detailed image, a DVD using red laser light or a DVD using blue? Explain why.

11) When using a telescope to look into outer space, which will give you more information and details, using the light waves or x-rays? Explain why.

12) Since the big Bang occurred 13.77 billion years ago, and its energy is lost over time, which category of waves would you expect to find its echo? Explain why.

13) Why do you think hospitals and restaurants use ultraviolet rays for sterilization?

14) Why do you think x-rays can see our bones?

How we use Microwaves

Directions:

You will need a **microwave**, two **microwave bowls**, **sand**, **water**, **oven mitts**, and a **digital thermometer** connected to an **interface** that is connected to a **computer** with **Logger Pro**. Microwaves cook food by flipping water molecules so fast that the friction creates heat. **Looking at the materials and lab we will be using, what are the safety precautions we should take to protect ourselves and materials during this investigation?**

1) Microwaves cook food by flipping water molecules so fast (about 2.45 billion times per second) that the friction creates heat. Pour dry sand into a microwave bowl and check the temperature of the sand with the temperature probe. Write this data in Data Table 1.
2) Place the dry sand into the microwave and run the microwave for one minute. Using oven mitts, take the bowl of sand out of the microwave and measure the temperature with the probe. Write this data in Data Table 1.
3) Subtract the two numbers and write down the change in temperature in Data Table 1.
4) Place some more sand into another bowl and add some water. Measure the temperature of the sand and water before putting it in the microwave and running it for one minute. Write this data in Data Table 1.
5) Using your oven mitts, take the bowl of sand out of the microwave and measure the temperature with the digital thermometer. Write this data in Data Table 1.
6) Subtract the two numbers and write down the change in temperature in Data Table 1.

Data Table 1

Type of Sand	Temperature before Microwave (^0C)	Temp after Microwave (^0C)	Change in Temperature
Dry Sand			
Wet Sand			

Questions:

1) Which bowl had a greater temperature change?

2) Why do you think this happened?

3) How does this explain why dried food stuck to the inside of the microwave does not get hot?

4) Will a microwave always cook food? Explain.

5) Why do you think microwaves interact with water this way? What else spins with electromagnetics?

Virtual Investigations that go with Electrons

ExploreLearning.com:

    Electron Configuration Gizmo

    Bohr Model: Introduction Gizmo

    Bohr Model of Hydrogen Gizmo

    Star Spectra Gizmo

    Photoelectric Effect Gizmo

Phet.colorado.edu:

    Balloons and Static Electricity

    Coulomb's Law

    Neon Lights and other Discharges

    Photoelectric Effect

    Lasers

    Simplifies MRI

    Radio waves and Electromagnetic Fields

    Blackbody Spectrum

    Conductivity

    Davisson-Germer: Electron Diffraction

    Semiconductors

    Stern-Gerlach Experiment

Physicsclassroom.com/Concept-Builders/Chemistry:

    Complete Electron Configuration

Unit 7: Periodic Table

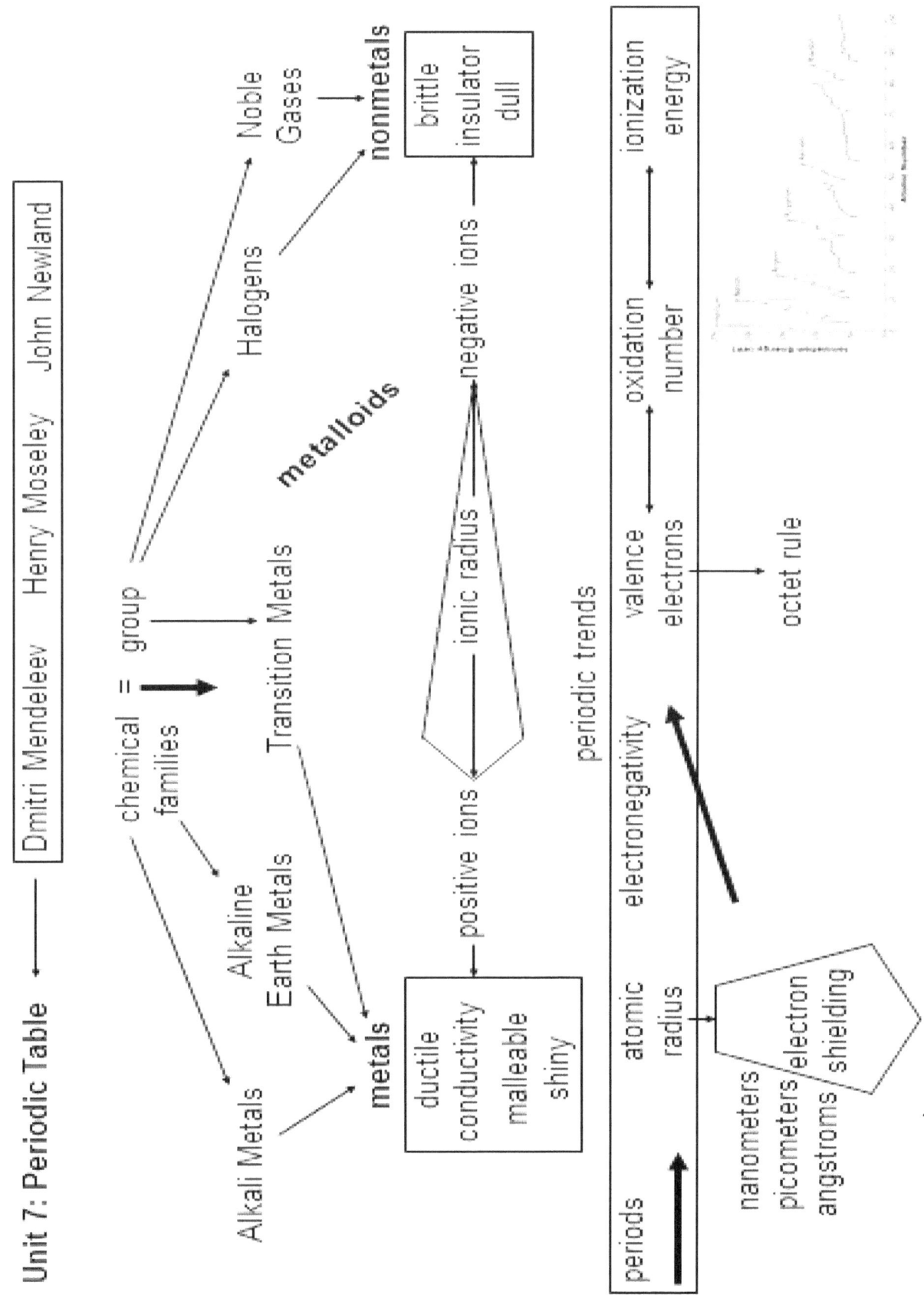

Flame Test

Directions:

You will need **safety goggles**, an **apron**, a box of **Q-tips**, a **Bunsen burner**, a **spark starter**, **8 test tubes**, a **test tube rack,** and **solutions** of **LiCl, $CaCl_2$, KCl, $CuCl_2$, $SrCl_2$, NaCl, $BaCl_2$. Looking at the materials and lab we will be using, what are the safety precautions we should take to protect ourselves and materials during the investigation?**

1) Make sure you have each of the solutions in the test tubes, and they are labeled with the solutions they have in them. The one tube unlabeled is the unknown with one of the solutions in them.
2) Turn on a Bunsen burner according to your teacher's instructions.
3) Take a Q-tip and dip it in the first test tube, soaking up the solution. Then slowly pass the soaked Q-tip through the lit Bunsen burner back and forth (without burning the Q-tip) until you see a color change in the flame. Write the flame color in Data Table 1.
4) Take a new Q-tip and repeat #3 for each solution, including the unknown.
5) Dispose of the Q-tips and clean up according to your teacher's instructions.

Data Table 1

Metal Ions	Li	Ca	K	Cu	Sr	Na	Ba	Unknown
Flame Color								

Because metallic elements are quickly excited, a flame test can detect which metal makes up that ion in the solution.

Questions:

1) Why do we not want to reuse a Q-tip in this investigation?

2) How do you think fireworks get their different colors?

3) What metallic property are we taking advantage of in this investigation?

4) What was the metallic ion in the unknown? How did you know?

Metal or Nonmetal

Directions:

You will need samples of **sulfur**, **charcoal**, **copper**, **aluminum foil**, **mechanical pencil lead** (graphite), a **battery pack**, **batteries**, and a **Christmas light** with the ends of the wires stripped of insulation. **Looking at the materials and lab we will be using, what are the safety precautions we should take to protect ourselves and materials during the investigation?**

1) **Metals** are shiny, malleable, ductile, and conduct heat and electricity well. **Nonmetals** are brittle, dull, and do not conduct heat and electricity well.
2) Observe the five samples and fill in Data Table 1 below on those materials. Use the battery pack with batteries and the Christmas light to see if the materials conduct enough electricity to light the Christmas light when put into a circuit.

Data Table 1

Sample	Shiny or Dull	Brittle or Malleable	Conduct Electricity?	Metal or Nonmetal?
Sulfur				
Charcoal				
Copper				
Aluminum Foil				
Mechanical Pencil Lead				

Questions:

1) What are the characteristics of metals?

2) What are the characteristics of nonmetals?

3) Look at the periodic table and find carbon. Why do you think graphite conducted electricity?

Simple Chemistry Investigations   Seven Sides Publishing

Finding the Period in the Periodic Table

Directions:

You will need a **periodic table**. Use the periodic table to fill in the Data Tables and answer the questions below.

Data Table 1

Element	Total # of Electrons	# of Level 1 Electrons	# of Level 2 Electrons	# of Level 3 Electrons
Magnesium				
Carbon				
Aluminum				
Sodium				
Florine				
Helium				
Oxygen				
Argon				
Silicon				
Nitrogen				

Questions for Data Table 1:

1) How many electrons can fill the first energy level?

2) How many elements do not fill that first energy level?

3) How many electrons fill the second energy level?

Data Table 2

Element	Energy Level of Outer Electrons	Located in period	Number of Outer Electrons	Located in Group
Magnesium				
Carbon				
Aluminum				
Sodium				
Fluorine				
Helium				
Oxygen				
Argon				
Silicon				
Nitrogen				

Questions Data Table 2:

1) How do you know an element's group?

2) Which elements have full outer shells?

3) What does the element's period tell us about the energy levels they have?

4) How does the periodic table seem to be organized?

Periodic Table Activity

Directions:

Using your **Periodic Table**, color the following with **colored pencils**:

1) Nonmetals are (yellow)
2) Metalloids are (green)
3) Metals are (blue)
4) Outline the alkali metals in (red)
5) Outline the alkaline earth metals in (black)
6) Outline the transition metals in (brown)
7) Outline the halogens in (blue)
8) Outline the noble gasses in (purple)

Add the following to the Periodic Table:

1) Atomic numbers
2) Group/family numbers
3) Periods/energy levels
4) Oxidation numbers
5) Label the Lanthanide series and Actinide series

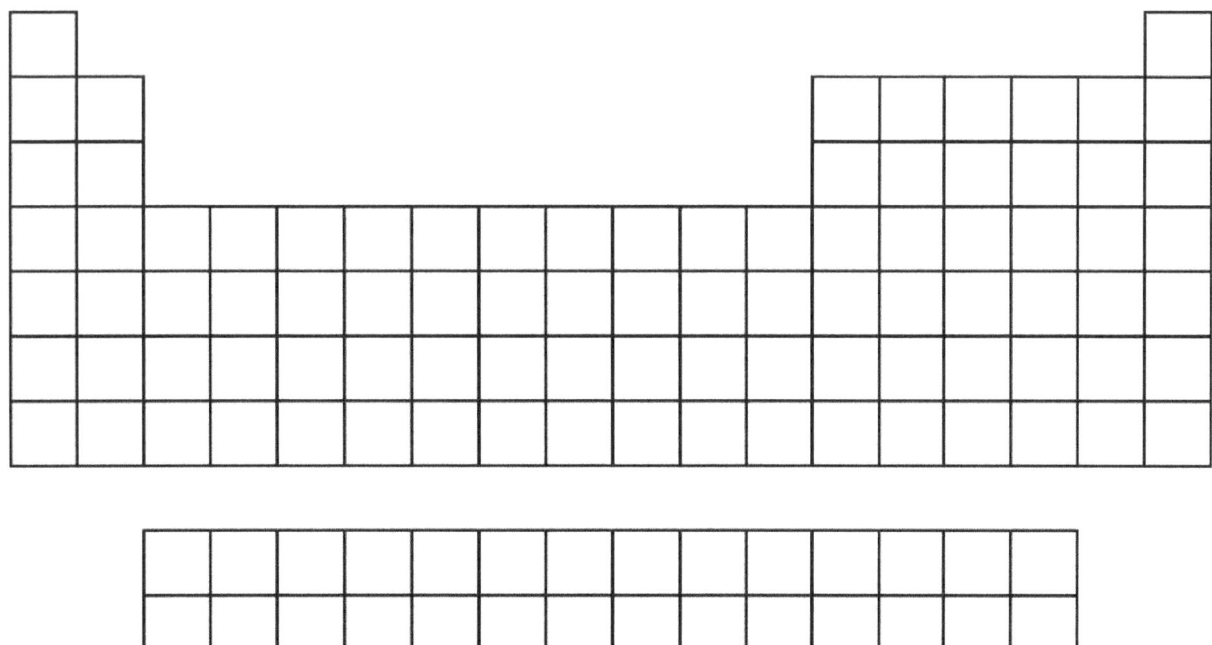

6) Use your textbook or internet to draw and label trends in:
   a. Atomic Radius
   b. Electronegativity
   c. Ionization energy

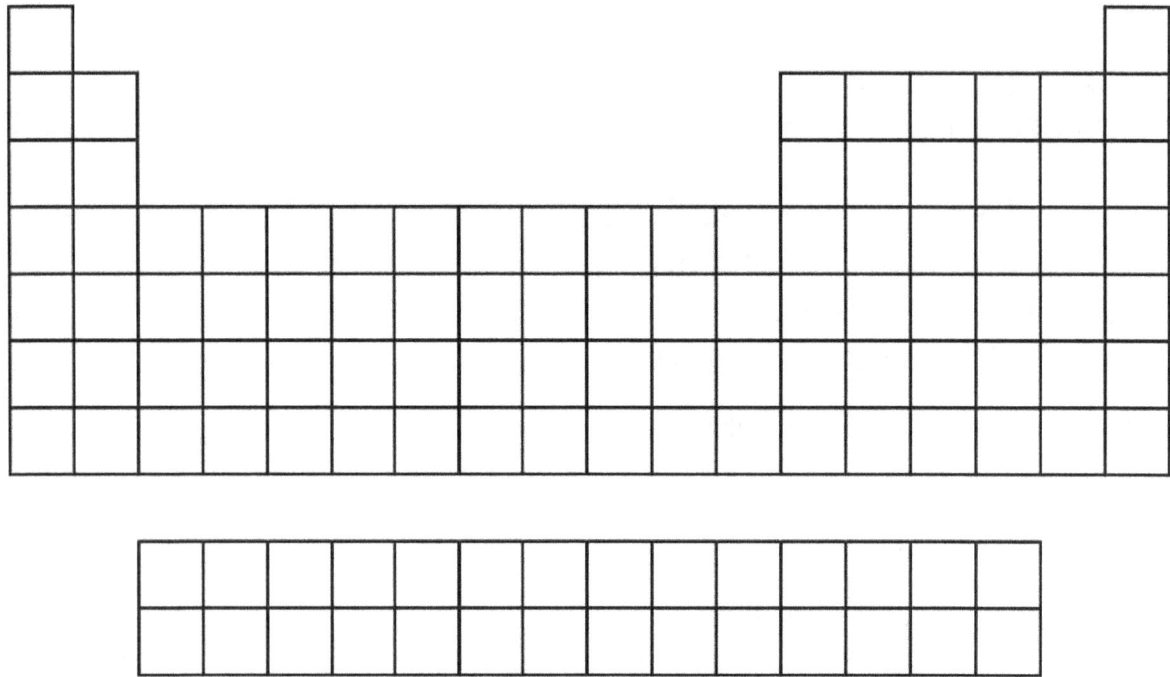

Questions:

1) What do groups have in common?

2) What do periods have in common?

3) What two things determine the radius of the atom?

4) Where are the most reactive metals?

5) Where are the most reactive nonmetals?

6) Which group/family does not react at all because they are full?

7) If you can't remember from previous investigations, research the internet to find these three things:
    a. What are the characteristics of metals?

    b. What are the characteristics of nonmetals?

    c. What are the characteristics of metalloids?

Simple Chemistry Investigations Seven Sides Publishing

Making Lewis Dot Structures for Elements

Directions:

Use the **Periodic Table** to help you construct Lewis dot structures showing valence electrons of the elements listed below. There are multiple ways to do this, so follow your teacher's example and recognize the patterns as you go through the periodic table. We will skip the transition metals since some things are happening there that get confusing.

1) H
2) He
3) Li
4) Be
5) B
6) C
7) N
8) O
9) F
10) Ne
11) Na
12) Mg
13) Al
14) Si
15) P
16) S
17) Cl
18) Ar
19) K
20) Ca
21) Ga
22) Ge
23) As
24) Se
25) Br
26) Kr
27) Rb
28) Sr
29) In
30) Sn

Making a Graphite Light Bulb

Directions:

You will need **.2 to .5 mm graphite mechanical pencil lead** (the thinner, the better because it causes more resistance), a **glass jar** with a **lid**, three **wires with alligator clips**, two **6 volt lantern batteries**, and **blue tac** (used to fix papers and posters to walls). **Looking at the materials and lab we will be using, what are the safety precautions we should take to protect ourselves and materials during the investigation?**

1) Take two pieces of the blue tac and fix them to the inside lid of the jar. Place an alligator clip from both wires to the blue tac holding the mouths up.
2) Take the mechanical pencil lead (graphite) and break a piece off big enough to fit in the two alligator clips and inside the jar's lid. Place the glass jar over the top of the lid covering the graphite and alligator clips.
3) Turn off the lights in the room. Take the other ends of the wires and place one on the "+" end of the first battery and the other wire on the "–" end of the second battery. To complete the circuit, take the third wire with alligator clips and place one clip on the "–" end of the first battery and the other clip on the "+" end of the second battery.
4) Make sure to disconnect the clips from the battery setup when you have finished.

Questions:

1) Why do you think the graphite lit up?

2) Look at the periodic table and find Carbon; this makes graphite. Why do you think this nonmetal was able to be used here?

3) Is Carbon a conductor or insulator? How do you know?

4) What do you see given off when you slow the flow of electrons in a part of the circuit?

5) How do you think an incandescent light bulb lights up when a current flows through it?

6) Explain how electrons and photons are involved with what is happening inside this light bulb.

7) Why did we cover the graphite with the glass jar?

Virtual Investigations that go with the Periodic Table

ExploreLearning.com:

    Element Builder Gizmo

    Periodic Trends Gizmo

Physicsclassroom.com/Concept-Builders/Chemistry:

    Metals, Nonmetals, and Metalloids

    Periodic Trends

Unit 8: Ionic and Metallic Bonding

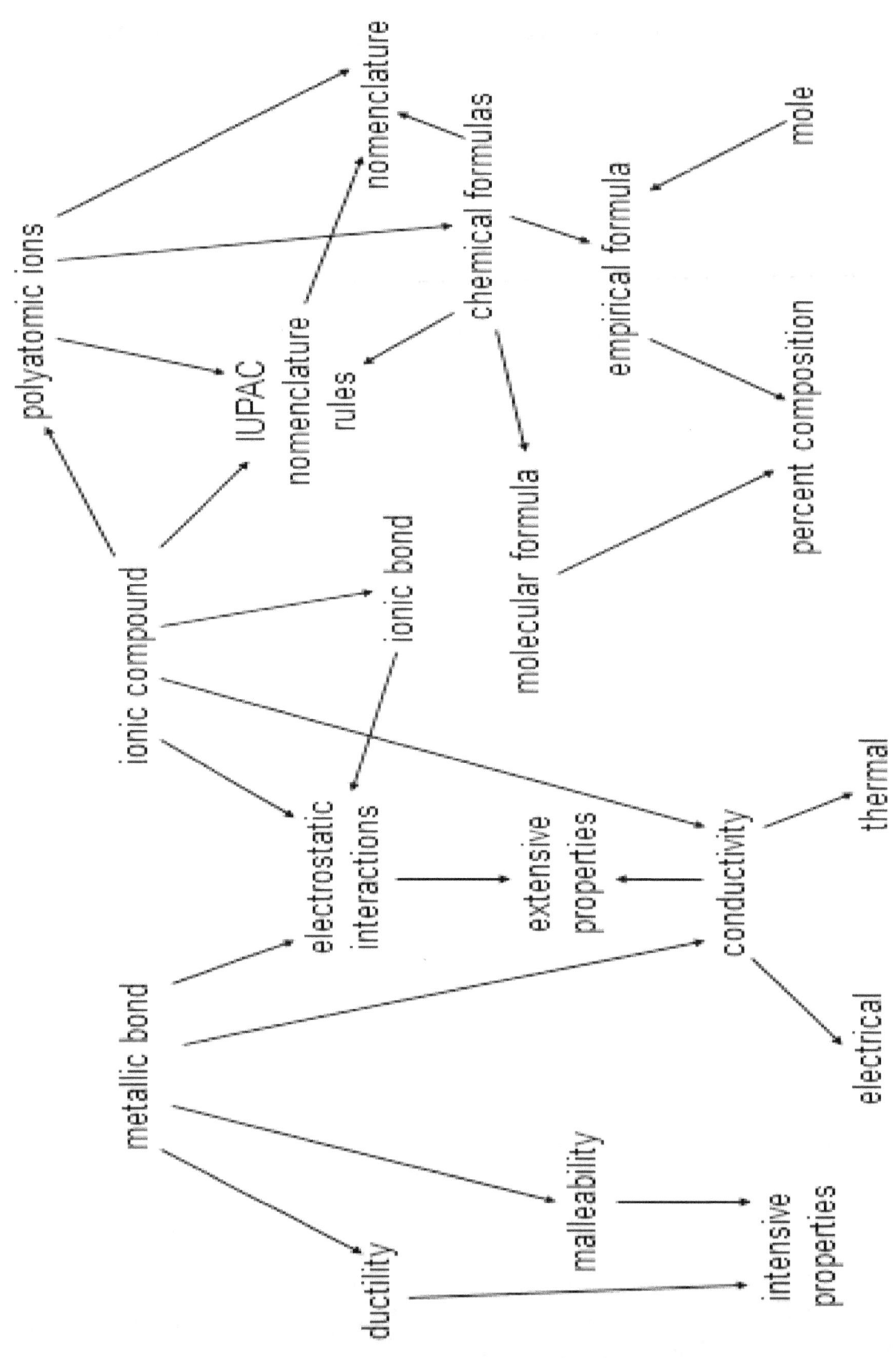

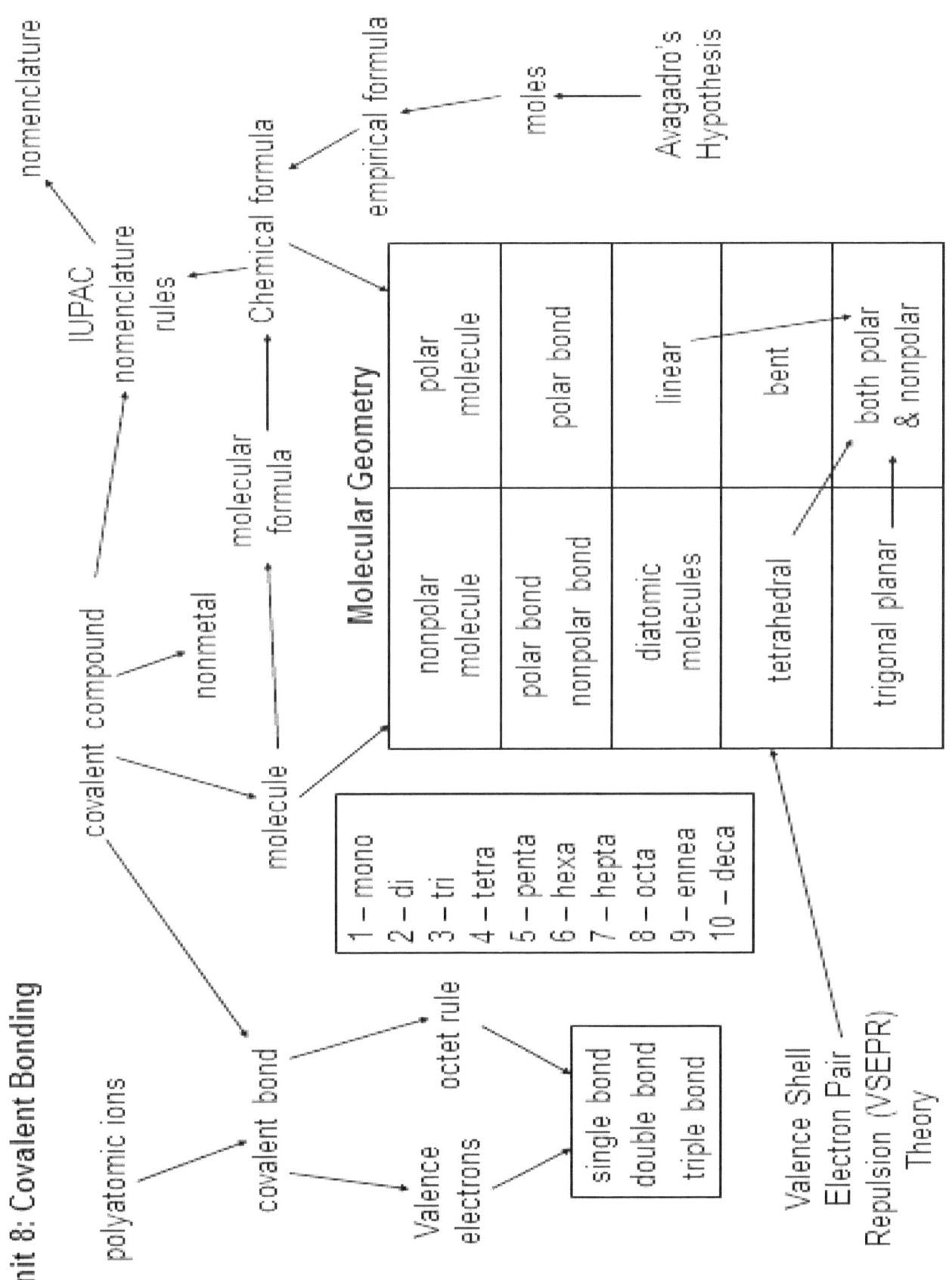

Simple Chemistry Investigations   Seven Sides Publishing

Ionic Bonding Models

Directions:

You will need a **molecular model kit** and a **Periodic Table**. Looking at the materials and lab we will be using, what are the safety precautions we should take to protect ourselves and materials during the investigation?

1) At the top of your periodic table, label it like this just below:

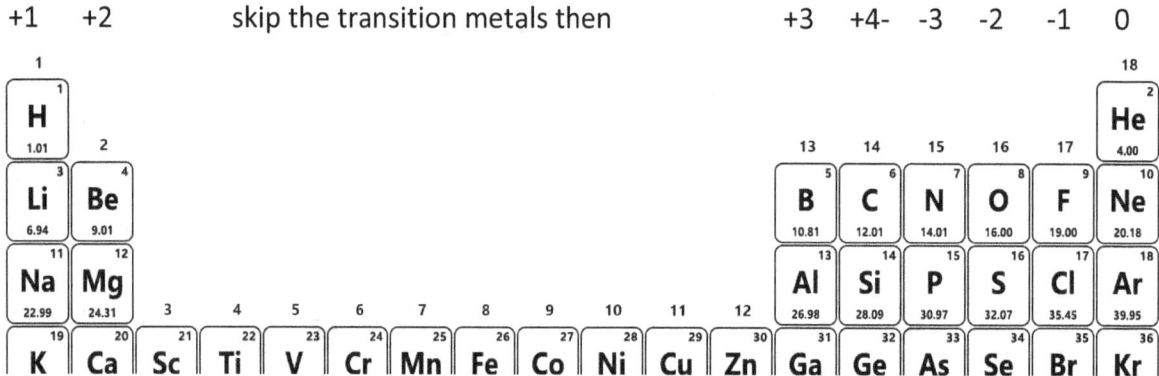

2) Different kits have different colors. In my kit, the:
   a. +1 (one-prong white) represents the Alkali Metals
   b. +2 (two-prong yellow) represents the Alkaline Earth Metals
   c. +3 (three-prong blue) represents the Boron Group
   d. +/- 4 (four-prong black) represents the Carbon Group
   e. -3 (three-prong red) represents the Nitrogen Group
   f. -2 (two-prong blue) represents the Oxygen Group
   g. -1 (one-prong green) represents the Halogens

3) The different pieces in #2 represent the elements in those groups. Put the following ionic compound together:

   KF MgI_2 BeS Na_2O $AlBr_3$ CH_3Cl NHF_2

4) Show the transfer of electrons for each compound you made in #3 using Lewis Dot Diagrams. Metals give electrons, and nonmetals take electrons.

   a. KF

b. MgI_2

c. BeS

d. Na_2O

e. $AlBr_3$

f. CH_3Cl

g. NHF_2

5) Name each compound.

   a. KF

   b. MgI_2

   c. BeS

   d. Na_2O

   e. $AlBr_3$

   f. CH_3Cl

   g. NHF_2

6) Calculate the percent composition of each element in each compound listed in #5.

Naming Ionic Compounds

Directions:

Use the **internet** and your **textbook** to find and use the IUPAC nomenclature to help you fill in the chart below, naming compounds, writing formulas, and counting elements and atoms.

Name	Formula	Number of Elements	Number of Atoms
Beryllium acetate			
Copper (I) arsenide			
Hydrochloric Acid			
Potassium chloride			
Magnesium phosphide			
Iron (II) bromide			
Barium oxide			
Ammonium hydroxide			
Sulfuric Acid			
Sodium hydroxide			
	CrF_2		
	Li_3PO_4		
	NaOH		
	Co_3N_2		
	$MgBr_2$		
	CaO		
	$(NH_4)_2SO_3$		
	$Al_2(SO_4)_3$		
	Na_3P		

Metallic Bond Research

Directions and Questions:

The task for you is to research the characteristics of metallic bonding. Use the **internet** as well as your **textbook** for your research. You must include the following information as part of your research:

1) Description of a metallic bond.

2) Explain the forces that hold the atoms and bonds together in metallic bonding.

3) Explain the properties of metallic elements caused by a metallic bonding.

4) Compare and contrast the characteristics of metallic bonds and ionic bonds.

5) Provide examples of compounds that represent ionic bond types and give the characteristics within the molecule.

6) What elements can be involved in metallic bonding?

7) Are there any exceptions?

8) Why are metals such good conductors of electricity?

Valence Shell Electron Pair Repulsion Theory

Directions and Questions:

Use the **internet** and your **textbook** to learn about VSEPR and answer the following questions.

1) What are valence electrons, and how are they used in this theory?

2) Why are electrons repulsive to each other?

3) How does the Lewis Structure affect the molecular geometry?

4) What are the basic simple molecular shapes, and how do they look?

5) How is this affected by double and triple bonds?

6) What is the role of nonbonding electrons?

7) Make notes or draw a chart organizing any trends you discovered.

Molecular Geometry

Directions:

You will need a **molecular model kit** and a **Periodic Table. Looking at the materials and lab we will be using, what are the safety precautions we should take to protect ourselves and materials during the investigation?**

1) At the top of your periodic table, label it like this just below:

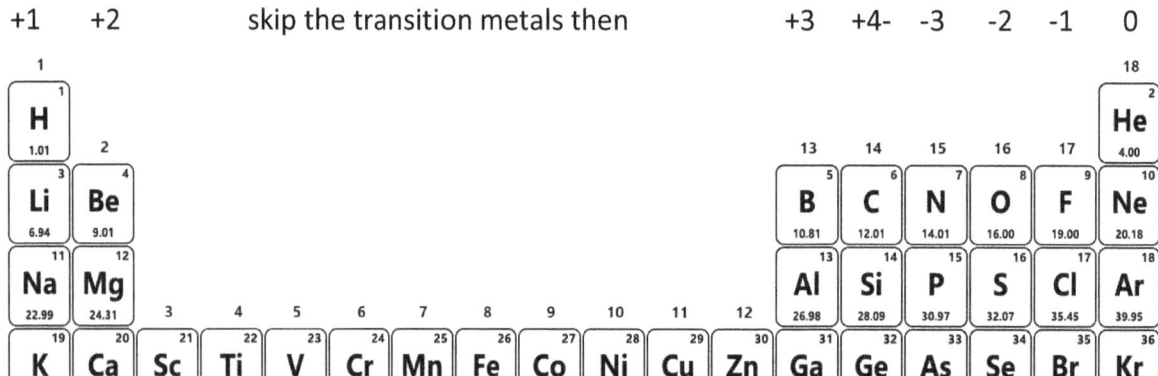

2) Different kits have different colors. In my kit, the:

   a. +1 (one-prong white) represents the Alkali Metals
   b. +2 (two-prong yellow) represents the Alkaline Earth Metals
   c. +3 (three-prong blue) represents the Boron Group
   d. +/- 4 (four-prong black) represents the Carbon Group
   e. -3 (three-prong red) represents the Nitrogen Group
   f. -2 (two-prong blue) represents the Oxygen Group
   g. -1 (one-prong green) represents the Halogens

3) The different pieces in #2 represent the elements in those groups. Put the following molecules together:

H_2O NH_3 $SiCl_4$ I_2 SCl_2 O_2 AsH_3

4) Draw the Lewis Dot Structures to show the sharing of electrons in covalent bonds in the chart below.
5) Using the Valence Shell Electron Pair Repulsion Theory and the Lewis Dot Structures, determine the molecules' geometric shapes in the chart below.

Molecule	Lewis Dot Structure	Shape of Molecule
H_2O		
NH_3		
$SiCl_4$		
I_2		
SCl_2		
O_2		
AsH_3		

6) Determine the name for each of these molecules.

   a. H_2O

   b. NH_3

   c. $SiCl_4$

   d. I_2

   e. SCl_2

   f. O_2

   g. AsH_3

7) Calculate the percent composition of each element in each compound listed in #6.

Building a Model of a Water Molecule

Directions:

You will need a **balloon**, a **molecular model kit**, and a **Periodic Table**. Looking at the materials and lab we will be using, what are the safety precautions we should take to protect ourselves and materials during the investigation?

1) At the top of your periodic table, label it like this just below:

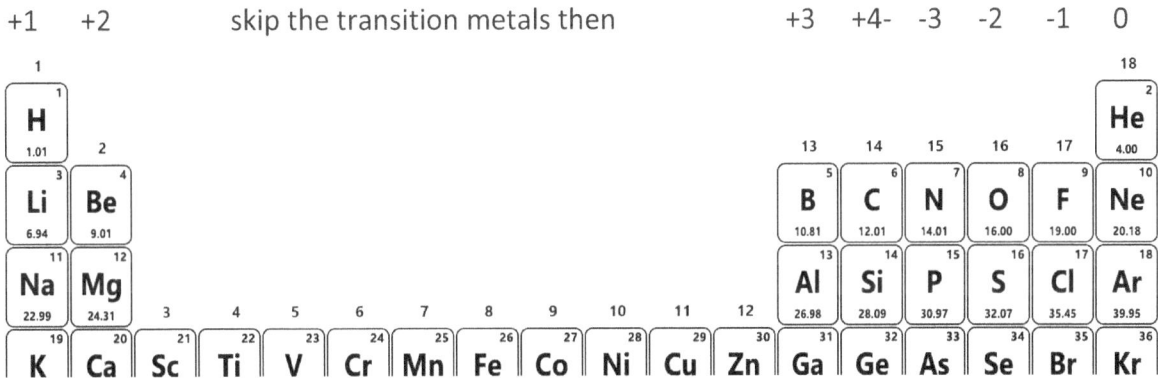

2) Different kits have different colors. In my kit, the:
   a. +1 (one-prong white) represents the Alkali Metals
   b. +2 (two-prong yellow) represents the Alkaline Earth Metals
   c. +3 (three-prong blue) represents the Boron Group
   d. +/- 4 (four-prong black) represents the Carbon Group
   e. -3 (three-prong red) represents the Nitrogen Group
   f. -2 (two-prong blue) represents the Oxygen Group
   g. -1 (one-prong green) represents the Halogens

3) Use the pieces to make two H_2O molecules. The hydrogen side of the molecule is slightly positive, and the oxygen side of the molecule is slightly negative making it polar like a magnet.

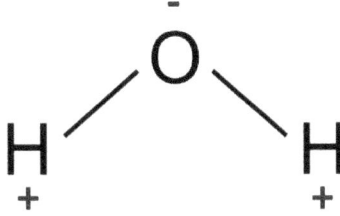

4) Because the water molecule has positive and negative ends, **ions** are attracted to the opposite charges on the water molecule. The positive ions are attracted to the oxygen side, and the negative ions are attracted to the hydrogen side. The same is true for other **polar molecules**; this is why ionic compounds and polar molecules like to dissolve in water. We call water the **universal solvent**.

5) Make a model of liquid water by taking your two water molecules and placing them next to each other where the oxygen of one is sitting between the two hydrogens of the other; this is how water molecules like to stick to each other. The positive ends are attracted to the negative ends, which is why water is **cohesive**.

6) When there are a bunch of them together (see the concept map), they have an equal pull on each other except for the ones on the surface; they are pulled slightly down because they have a slight charge above them. After all, there are no other molecules above them; this is why water has **surface tension**.

7) Since water is charged on both ends, it is also attracted to surfaces; like a balloon with a static charge is attracted to a sweater or a wall. These charges are why we see water clinging to the sides of cold cans or glasses of ice tea. This phenomenon is called **adhesion**.
    a. You can model this by taking an inflated balloon, rubbing it on your hair to steal some electrons, and then sticking it to a shirt or wall.

8) You can make a model of ice (solid water) by flipping one of your water molecules and facing the hydrogen ends toward each other. When water gets cold, the molecule's charge is not as strong, and the opposite ends are not attracted to each other anymore, so the oxygen atoms come together and share electrons with each other bonding them together. This action gives the molecule more space inside it, which is why ice floats in liquid water.

Question:

1) Why do you think life depends so much on water?

How does Rain Form?

Directions and Questions:

You will need a glass or beaker of ice water. **Looking at the materials and lab we will be using, what are the safety precautions we should take to protect ourselves and materials during the investigation?**

1) Why do you see water forming on the outside of the glass?

2) How is this like water forming droplets in the sky, making clouds and rain?

   a. What is the difference between clouds and rain?

3) How is a liquid different from a gas allowing this to happen with water?

4) What allows the water to stick together on the glass?

5) What holds the water drops to the glass?

6) How are **cohesion**, **adhesion**, and **surface tension** involved?

7) If an animal was small enough, could it walk on water? Explain why

Naming Covalent Compounds

Directions:

Use the **internet** and your **textbook** to find and use the IUPAC nomenclature to help you fill in the chart below, naming compounds, writing formulas, and counting elements and atoms.

Name	Formula	Number of Elements	Number of Atoms
Nitrogen tribromide			
Hexaboron silicide			
Chlorine dioxide			
Hydrogen iodide			
Iodine pentafluoride			
Dinitrogen trioxide			
Nitrogen trihydride			
Phosphorous triiodide			
Dihydrogen monoxide			
Diphosphourous pentoxide			
	P_4S_5		
	O_2		
	SF_6		
	Si_2Br_6		
	CH_4		
	B_2Si		
	NF_3		
	H_2O		
	N_2O_5		

Empirical and Molecular Formulas

Directions and Questions:

1) The empirical formula is CH_2O. What is the molecular formula if its molar mass is 180 g/mol?

2) The empirical formula is H_2O. What is the molecular formula if its molar mass is 18 g/mol?

3) Calculate the empirical formula if the compound is 67.6% Hg, 10.8% S, and 21.6% O.

   a. What would be the molecular formula if its molar mass is 745.77 g/mol?

4) Find the empirical and molecular formulas if a compound has 40% C, 6.6% H, and the rest O with a molar mass of 120 g/mol.

5) What are the empirical and molecular formulas for a compound with 18.7% Li, 16.3% C, and 65% O and a 148 g/mol molar mass?

Breaking Bonds

Directions and Questions:

You will need **safety goggles**, **salt**, **sugar**, and a **hotplate. Looking at the materials and lab we will be using, what are the safety precautions we should take to protect ourselves and materials during the investigation?**

1) We are going to see which bond is stronger: ionic or covalent. Salt is ionically bonded, and sugar is covalently bonded. Place a pinch of each on a hotplate and turn the temperature on high. The first one to burn has the weakest bond. Once one of them melts or burns, turn off the burner.

2) Which one burned first?

3) Why do you think this bond is weaker?

The Conductivity of Electrolyte Mixtures

Directions:

You will need a **battery pack**, **batteries**, **solutions** listed in Data Table 1, a **conductivity probe** attached to an **interface** connected to a **computer** with **Logger Pro**, and a **Christmas light** to check the solutions shown in Data Table 1 for conductivity. **Looking at the materials and lab we will be using, what are the safety precautions we should take to protect ourselves and materials during the investigation?**

1) Connect one wire from the Christmas light to the battery pack with batteries in it. Put the other wires from the battery pack and Christmas light together to see if the light bulb lights. Then dip the free wires into the solutions in the chart below to see if the lightbulb lights. If it lights, the solution conducts electricity; if it does not, it does not conduct electricity.
2) Set your conductivity probe to 0-2000 µS/cm). Then measure the conductivity of the substances in the table below, rinsing the probe between each solution.

Data Table 1

Mixture	Light On or Off	Conductivity (µS/cm)
Tap Water (covalent)		
Salt Water (ionic)		
Soap (covalent)		
Orange Juice (both)		
Sugar Water (covalent)		
Soda (both)		
Vinegar (ionic)		
Milk (both)		
Gatorade or Powerade		

Questions:

1) What pattern do you see about which mixtures conducted electricity or not?

2) Do covalently bonded substances conduct electricity?

3) Do ionically bonded substances conduct electricity?

4) Why do you think this type of bond conducts electricity?

5) Gatorade and Powerade are promoted to replenish electrolytes; does the data show they have many in them?

6) Why do we need to replenish electrolytes?

Simple Chemistry Investigations

Polarity of Water

Directions and Questions:

You will need an **inflated balloon** and a small **stream of water. Looking at the materials and lab we will be using, what are the safety precautions we should take to protect ourselves and materials during the investigation?**

1) Rub the balloon on your hair (if it does not have hair spray or moose).
2) Bring the balloon you rubbed on your hair near a small stream of water coming out of the faucet.
3) What do you observe happening?

4) The balloon is charged and acts as a magnet. Why do you think the stream of water is also acting as a magnet when it is covalently bonded?

5) What does polarity mean?

6) Why does this show water is polar?

Checking Polarity

Directions:

You will need two **pennies** tails up on a **paper towel**, two **pipettes** or **eyedroppers**, **water**, and **rubbing alcohol. Looking at the materials and lab we will be using, what are the safety precautions we should take to protect ourselves and materials during the investigation?**

1) With a pipet or eyedropper, slowly place drops of water on one penny. Count how many drops you could put on the penny before it spilled off. Write that number in Data Table 1 below.
2) With a different pipet/eyedropper, slowly place drops of rubbing alcohol on the second penny. Count how many drops you could put on the penny before it spilled off. Write that number in Data Table 1.

Data Table 1

Substance Dropped	Number of Drops
Water	
Alcohol	

Questions:

1) Which substance was able to have the most drops put onto the penny?

2) Which substance had more polarity?

3) How did the polarity work to allow the molecules to stick together?

4) How does water defy gravity by moving up the stems of plants without pulling them apart?

Virtual Investigations that go with Bonding

ExploreLearning.com:

    Ionic Bonds Gizmo

    Covalent Bonds Gizmo

    Polarity and Intermolecular Forces Gizmo

    Melting Points Gizmo

    Sticky Molecules Gizmo

Phet.colorado.edu:

    Molecular Polarity

    Molecule Shapes

    Molecule Shapes: Basics

    Atomic Interactions

    Build a Molecule

Physicsclassroom.com/Concept-Builders/Chemistry:

    Ionic Bonds

    Lewis Structures

    Valence Shell Electron Pair Repulsion Theory

    Particles…Words…Formulas

    Names and Formulas 1

    Names and Formulas 2

    Formulas and Atom Counting

    Dissociation

Unit 9: Energy in Chemical Changes

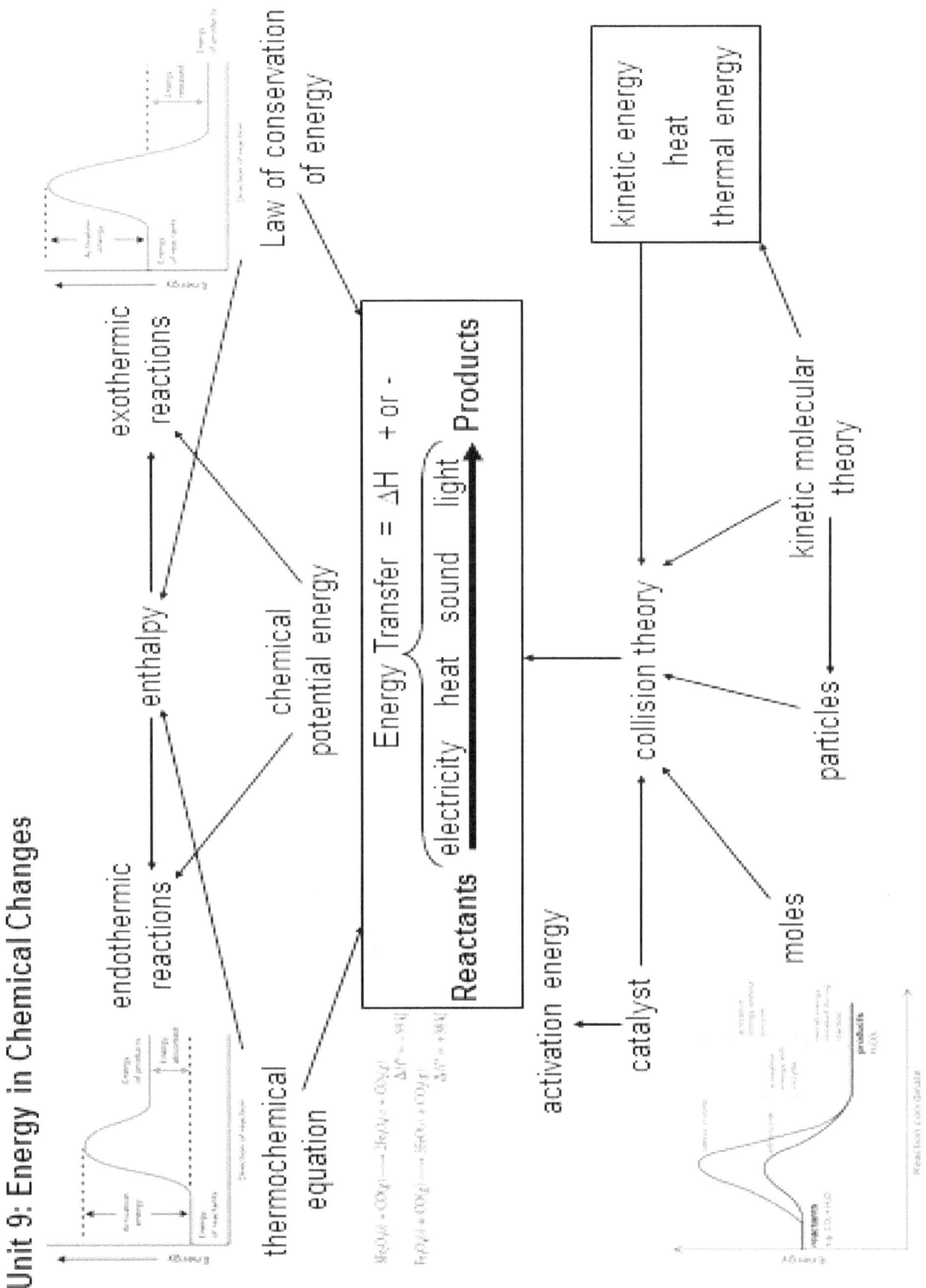

Interpreting Activation Energy for a Reaction

Directions:

Use the graph below that shows the difference in activation energy for a chemical reaction to answer the following questions.

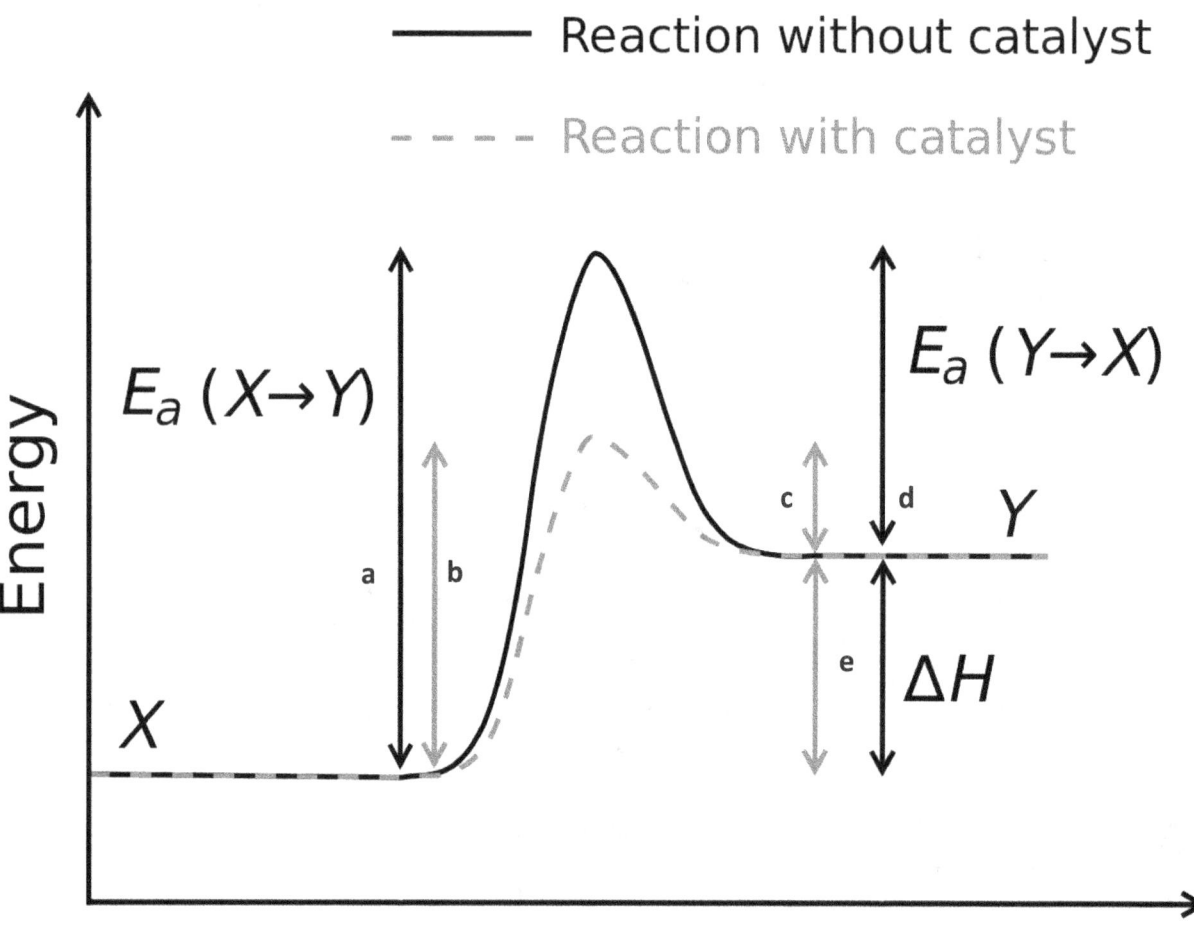

Questions:

1) X is the energy in the reactant(s), and Y is the energy in the product(s). How does the energy change from before the reaction to after the reaction?

2) Is this an endothermic or exothermic reaction?

3) Would heat be released or absorbed?

4) The activation energy is the energy required for the reaction to occur. Which has lower activation energy, the reaction by itself or the reaction with a catalyst?

5) How does this explain why catalysts speed up reactions?

6) Which letter on the graph, (a) (b) (c) (d) or (e), is the enthalpy of the reaction?

7) Which letter is the activation energy for the reaction with a catalyst?

8) Which letter is the activation energy for the reaction without a catalyst?

9) Which letter is the activation energy for the reverse reaction with a catalyst?

10) Which letter is the activation energy for the reverse reaction without a catalyst?

11) Is the reverse reaction of this graph endothermic or exothermic?

12) Would the reverse reaction take in or give off energy?

Temperature and Reaction Rates

Directions:

You will need **safety goggles**, a **beaker** of **cold water** cooled by a **refrigerator**, a **beaker** of **warm water** heated by a **hotplate**, and two **Alka-Seltzer tablets**. **Looking at the materials and lab we will be using, what are the safety precautions we should take to protect ourselves and materials during the investigation?**

1) Heat one beaker with a hotplate to the point just before it boils.

2) Then take the other beaker out of the refrigerator.

3) At the same time, add an Alka-Seltzer tablet to each of the beakers and see which tablet finishes reacting first.

Questions:

1) Which water made the tablet react the fastest?

2) Why do you think this happened?

Observing a Catalyst

Directions and Questions:

You will need **safety goggles**, an **apron**, a small bottle of **soda** (diet has more fizz), and a **Mentos tablet**. Looking at the materials and lab we will be using, what are the safety precautions we should take to protect ourselves and materials during the investigation?

1) Open the bottle of soda. What do you observe happening?

2) Now drop the Mentos (the catalyst) into the bottle of soda. What do you observe now?

3) This reaction converted all the carbonic acid in the soda to carbon dioxide and water. The Mentos catalyst will keep working until all the carbonic acid is gone without using up the catalyst. Try to devise an experiment showing the catalyst does not go away.

Endothermic and Exothermic Reactions

Directions and Questions:

You will need **safety goggles**, a **beaker**, **water**, **Epsom salt**, **Borax**, **laundry detergent**, **baking soda**, **vinegar**, **matches**, and a **temperature probe** attached to an **interface** connected to a **computer** with **Logger Pro**. Looking at the materials and lab we will be using, what are the safety precautions we should take to protect ourselves and materials during the investigation?

1) **Endothermic Reactions** absorb energy into the reactions making the measured temperature drop. **Exothermic reactions** release energy into the environment making the measured temperature go up. We will conduct a few simple reactions and measure with a temperature probe whether the temperature goes up or down and thus classify those reactions as endothermic or exothermic.

2) Put a temperature probe into a beaker of water. Once the temperature stabilizes, add a small spoon full of Epsom salt to it. Gently stir the mixture slowly, watching the temperature readings on Logger Pro. What happened to the temperature during this reaction?

   a. Was this an endothermic or exothermic reaction? How do you know?

3) Rinse everything off and put water into your beaker again. Put a temperature probe into the beaker of water. Once the temperature stabilizes, add a small spoonful of borax to it. Gently stir the mixture with the temperature probe as it reacts, watching the Logger Pro temperature reading. What happened to the temperature during this reaction?

   a. Was this an endothermic or exothermic reaction? How do you know?

4) Rinse everything off and put water into your beaker again. Put a temperature probe into the water. Once the temperature stabilizes, add laundry detergent. Gently stir the

mixture as it reacts, watching the temperature reading on the Logger Pro. What happened to the temperature during the reaction?

   a. Was this an endothermic or exothermic reaction? How do you know?

5) Rinse everything off and put water into your beaker again. Put a temperature probe into the water. Once the temperature stabilizes, add baking soda to it. Gently stir the mixture as it reacts, watching the temperature reading on the Logger Pro. What happened to the temperature during the reaction?

   a. Was this an endothermic or exothermic reaction? How do you know?

6) Rinse everything off and put vinegar into your beaker this time. Put the temperature probe into the vinegar. Once the temperature stabilizes, add baking soda to it. Gently stir the mixture as it reacts, watching the temperature readings on the Logger Pro. What happened to the temperature during the reaction?

   a. Was this an endothermic or exothermic reaction? How do you know?

7) Light a match and watch the reaction take place with the resulting fire. Is this an endothermic or exothermic reaction? How do you know?

8) When fireworks go off, is this an endothermic or exothermic reaction? How do you know?

9) Instant cold packs are made up of two bags, one inside the other. One bag contains water, while the other holds a chemical like calcium ammonium nitrate. When you hit it and shake it, the bag on the inside breaks mixing the two contents causing the temperature to drop; is this an endothermic or exothermic reaction? How do you know?

Endothermic and Exothermic Models

Directions:

Use the **internet**, your **textbook**, and knowledge from this unit to develop a model to illustrate the release and absorption of energy in a chemical reaction for both endothermic and exothermic reactions. In each reaction, show: where energy is released and absorbed by the system, the change in enthalpy, and how to read the reverse reaction. Describe your models below.

Endothermic Reaction:

Exothermic Reaction:

Virtual Investigations that go with Energy in Chemical Changes

ExploreLearning.com:

- Collision Theory Gizmo
- Equilibrium and Concentration Gizmo
- Reaction Energy Gizmo
- Feel the Heat Gizmo

Phet.colorado.net:

- Microwave
- Molecules and Light
- Reactions and Rates
- Reversible Reactions
- The Greenhouse Effect

Physicsclassroom.com/Concept-Builders/Chemistry:

- Which One Doesn't Belong? – Energy and Chemical Reactions
- Collision Model of Reaction Rates
- Le Chatelier's Principle
- Pressure Concepts

Unit 10: Classification of Chemical Reactions

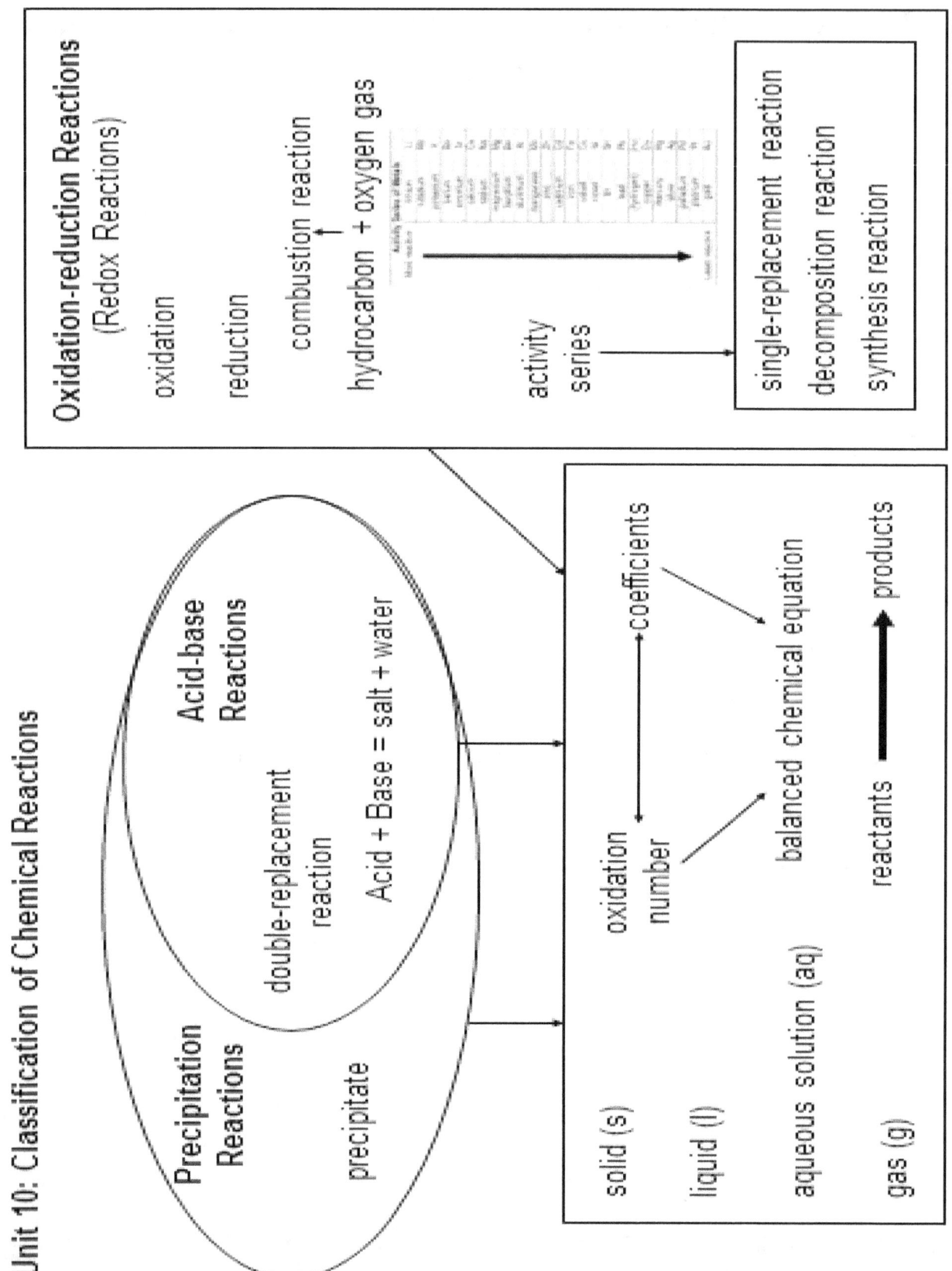

Home Chemistry

Directions and Questions:

You will need **safety goggles**, **liver**, **hydrogen peroxide**, three small **cups**, **water**, **vinegar**, and **baking soda**. **Looking at the materials and lab we will be using, what are the safety precautions we should take to protect ourselves and materials during the investigation?**

1) All living things produce an enzyme called catalase. Catalase helps break down hydrogen peroxide into water and oxygen. Place a piece of liver in a cup and pour some hydrogen peroxide we use as an antiseptic on it. What do you observe?

   a. Write a balanced equation for the reaction.

   b. What type of reaction is this?

2) One of the most useful natural gasses we use as fuel is methane (CH_4). It combines with oxygen in a combustion reaction to create heat, carbon dioxide, and water. Write a balanced equation for the reaction.

   a. What type of reaction is this?

3) Place baking soda (NaHCO₃) and vinegar (CH₃COOH) together in a cup and watch the reaction. What do you observe?

   a. Write a balanced equation for this reaction.

   b. What type of reaction is this?

4) What do you think we use baking soda for when we cook?

Types of Chemical Reactions

Equipment and Safety:

You will need **safety goggles**, an **apron**, a **small aluminum pan**, a **digital scale**, **steel wool**, **baking soda**, **matches**, a **copper sulfate solution** in a **beaker**, a **nail or screw**, a **lighter**, a **small beaker**, **beaker tongs**, a **test tube**, and a **test tube holder**. Looking at the materials and lab we will be using, what are the safety precautions we should take to protect ourselves and materials during the investigation?

Prep for Reaction 3

1) Take the <u>**nail or screw**</u> and write down how it appears now.

2) Stand it up in the small amount of <u>**copper sulfate**</u> and come back and look at it after the other reactions are done.

Reaction 1

3) Place the aluminum pan on the digital scale and zero it out.
4) Take some <u>**steel wool**</u> and place it in a pan on a digital scale. What is the mass of the steel wool?

5) What is the color of the steel wool before burning?

6) Take the lighter and light the steel wool. Place the lighter's flame into the steel wool and observe the reaction. What do you see happening during the reaction?

7) Notice the color change. What is the color of the burned steel wool?

8) What is the mass of the burnt steel wool?

9) Why do you think the mass changed?

10) This reaction was a combustion reaction where the iron in the steel wool combined with oxygen in the air, using fire, forming the iron oxide. Write a word equation of the reaction.

11) If I were to tell you this iron was iron III, write a balanced equation for the chemical reaction.

12) What type of reaction is this combustion reaction?

Reaction 2

13) Take some **baking soda** ($NaHCO_3$) and place it in a test tube; make sure you hold the test tube with the test tube holder. Take the lighter and heat the test tube with the baking soda. What do you see forming on the inside of the glass of the test tube?

14) Now light a match and place it inside the mouth of the test tube. What happens to the flame?

15) What color is the solid at the bottom of the test tube?

16) What you saw was water form on the inside of the test tube; and carbon dioxide gas form that snuffed the flame of the match out. Sodium carbonate is what is left. Write a word equation for the reaction.

17) Write a balanced chemical equation for the reaction.

18) What type of reaction was this?

Reaction 3

19) Go back to your **nail or screw** and carefully pull it out of the liquid. How does it look now?

20) The iron and the copper switched places. The iron is now combined with the sulfate, and the copper is by itself. Write a word equation for the reaction.

21) Write a balanced equation for this reaction.

22) What type of reaction was this?

Acid-Base Reaction

23) Write a word equation for hydrochloric acid mixing with sodium hydroxide that makes water and table salt.

24) Write a balanced equation for this reaction.

Removing Carbon from Sugar

Directions and Questions:

Perform this reaction under a **fume hood** or go **outside**. Some unhealthy gasses and heat are produced from this reaction, so once it has started, stand back and watch but do not touch. You will need **safety goggles**, an **apron**, **sugar**, **baking soda**, **lighter fluid**, a **ceramic bowl with sand**, and a **long-necked lighter**. Looking at the materials and lab we will be using, what are the safety precautions we should take to protect ourselves and materials during the investigation?

1) Now take the ceramic bowl of sand and have your teacher spray some lighter fluid on it. Place a mixture of 40 g sugar and 10 g baking soda on it. Make sure there is nothing near the bowl. Have a fire extinguisher ready if the fire gets out of hand. Then with a long-neck lighter, light it on fire. What do you see happening?

2) The sugar reacts with the oxygen producing carbon dioxide and water. When the oxygen runs out, the sugar breaks down to solid carbon (what looks like a black snake) and water vapor. The baking soda changed to carbon dioxide, water, and sodium carbonate (this is what is stealing the oxygen to help the sugar create a black snake). Write the word equations for the reactions taking place.

3) Write balanced chemical reactions for the reactions taking place.

4) How would you classify these reactions?

5) Was this an endothermic or exothermic reaction?

6) Since this is burning sugar like life does to live, is respiration an endothermic or exothermic reaction?

7) Wait until it cools before touching it. The black snake will stain whatever touches it, so be careful as you dispose of it using your teacher's instructions.

Nonrenewable Resources Chart

Directions:

Use the **internet** and your **textbook** to research and fill out this chart on nonrenewable energy resources.

Type	How do we obtain and transport it	How we use it	How it affects the environment
Petroleum			
Coal			
Natural Gas			
Nuclear			

Questions:

1) What are the advantages of using petroleum?

2) What are the disadvantages of using petroleum?

3) What type of reaction goes into burning petroleum products for energy?

   a. Show an example of a balanced equation of a reaction using a petroleum product for energy.

   b. Is it an endothermic or exothermic reaction?

   c. Are there any products from this reaction that are harmful to the environment? If so, how are they harmful?

4) What are the advantages of using coal?

5) What are the disadvantages of using coal?

6) What type of reaction goes into burning coal for energy?

   a. Show an example of a balanced equation of a reaction for burning coal.

   b. Is it an endothermic or exothermic reaction?

   c. Are there any products from this reaction that are harmful to the environment? If so, how are they harmful?

7) What are the advantages of using natural gas?

8) What are the disadvantages of using natural gas?

9) What type of reaction goes into burning natural gas?

   a. Show an example of a balanced equation of a reaction burning natural gas for energy.

   b. Is it an endothermic or exothermic reaction?

c. Are there any of the products from this reaction that are harmful to the environment? If so, how are they harmful?

10) What are the advantages of using nuclear energy?

11) What are the disadvantages of using nuclear energy?

12) Show an example of a balanced equation when nuclear energy is used to make electricity.

   a. What type of reaction do you think this is?

   b. Is this reaction endothermic or exothermic?

   c. Are there any products from nuclear reactions that are harmful to the environment? If so, how are they harmful?

13) Describe some careers involved with the exploration, extraction, production, and disposal of these resources.

Renewable Resources Chart

Directions:

Use the **internet** and your **textbook** to research and fill out this chart on renewable energy resources.

Type	How we obtain and transport it	How we use it	How it affects the environment
Wind			
Solar			
Hydroelectric			
Geothermal			

Questions:

1) What are the advantages of using wind energy?

2) What are the disadvantages of using wind energy?

3) What are the advantages of using solar energy?

4) What are the disadvantages of using solar energy?

5) What are the advantages of using hydroelectric energy?

6) What are the disadvantages of using hydroelectric energy?

7) What are the advantages of using geothermal energy?

8) What are the disadvantages of using geothermal?

9) Describe some careers involved with the production of energy using these resources.

Oxidation-Reduction Titration

Directions:

You will need **safety goggles**, an **apron**, a **graduated cylinder**, a **250 mL beaker**, **bleach**, **hydrogen peroxide**, a **ring stand**, **clamps** to hold a **50 mL burette**, an **Oxidation-Reduction Potential Sensor** attached to an **interface** connected to a **computer** with **Logger Pro**, and a **stirring rod** or **magnetic stirrer**. Looking at the materials and lab we will be using, what are the safety precautions we should take to protect ourselves and materials during the investigation?

1) Fill the 50 mL burette with hydrogen peroxide. What is the concentration of hydrogen peroxide?
2) Measure 10 mL of bleach with the graduated cylinder. Place the bleach in a 250 mL beaker. Add 100 mL of distilled water to dilute the solution.
3) Place the bleach solution under the burette and ORP Sensor.
4) Start the data collection for the ORP Sensor. Set up data collection for events with entry mode. Add the magnetic stirrer or stir with a stirring rod. When the reading stabilizes, select "keep" and enter "0", then select "OK."
5) Add 1 mL of hydrogen peroxide (keep stirring), and when the reading stabilizes, select "keep" and enter the burette reading, then click "OK."
6) Keep repeating step #5 until the potential value remains constant. Estimate the volume of hydrogen peroxide it took to reach this point.
7) The reaction that just occurred was: $NaOCl + H_2O_2 \longrightarrow NaCl + H_2O + O_2$
8) Calculate the number of moles of hydrogen peroxide used.

9) Calculate the number of moles of bleach in your sample.

10) Using your measured volume of bleach, now calculate the molarity of the bleach in your sample.

Using Activity Series to Predict Reactions

Directions:

Use the Activity Series to predict whether a reaction will occur or not. This only works in Single Replacement (single displacement) Reactions. If the element is listed above the (+) ion in the compound, then it will react and replace that (+) ion. If it is below that (+) ion on the list, it will not react. If the reaction will occur, write the product on the right side of the arrow and balance the equation. If the reaction does not occur, write no reaction or (NR) to the right of the arrow.

Activity Series (Increasing Activity ↑)
Lithium
Potassium
Boron
Calcium
Sodium
Magnesium
Aluminum
Manganese
Zinc
Chromium
Iron
Cobalt
Nickle
Tin
Lead
(Hydrogen)
Copper
Mercury
Silver
Platinum
Gold

$Al + HCl \rightarrow$

$Ag + Cu(NO_3)_2 \rightarrow$

$Li + NaCl \rightarrow$

$Pb + LiF \rightarrow$

$Al_2(SO_4)_3 + Fe \rightarrow$

$Ag_2O + Zn \rightarrow$

$Ca + MnSO_4 \rightarrow$

$Sn + LiF \rightarrow$

$B_2(CO_3)_3 + Cr \rightarrow$

$MgI_2 + K \rightarrow$

The same rules apply to the Halogens and (−) ions

Halogens
Fluorine
Chlorine
Bromine
Iodine

$F_2 + NaCl \rightarrow$

$Br_2 + HF \rightarrow$

Virtual Investigations that go with Classification of Chemical Reactions

ExploreLearning.com:

    Chemical Equations Gizmo

    Balancing Chemical Equations Gizmo

    Chemical Changes Gizmo

Phet.colorado.edu:

    Balancing Chemical Equations

    Reactants Products and Leftovers

Physicsclassroom.com/Concept-Builders/Chemistry:

    Balancing Chemical Equations

    Chemical Reaction Types

    Writing Balanced Chemical Equations

    Precipitation Reactions

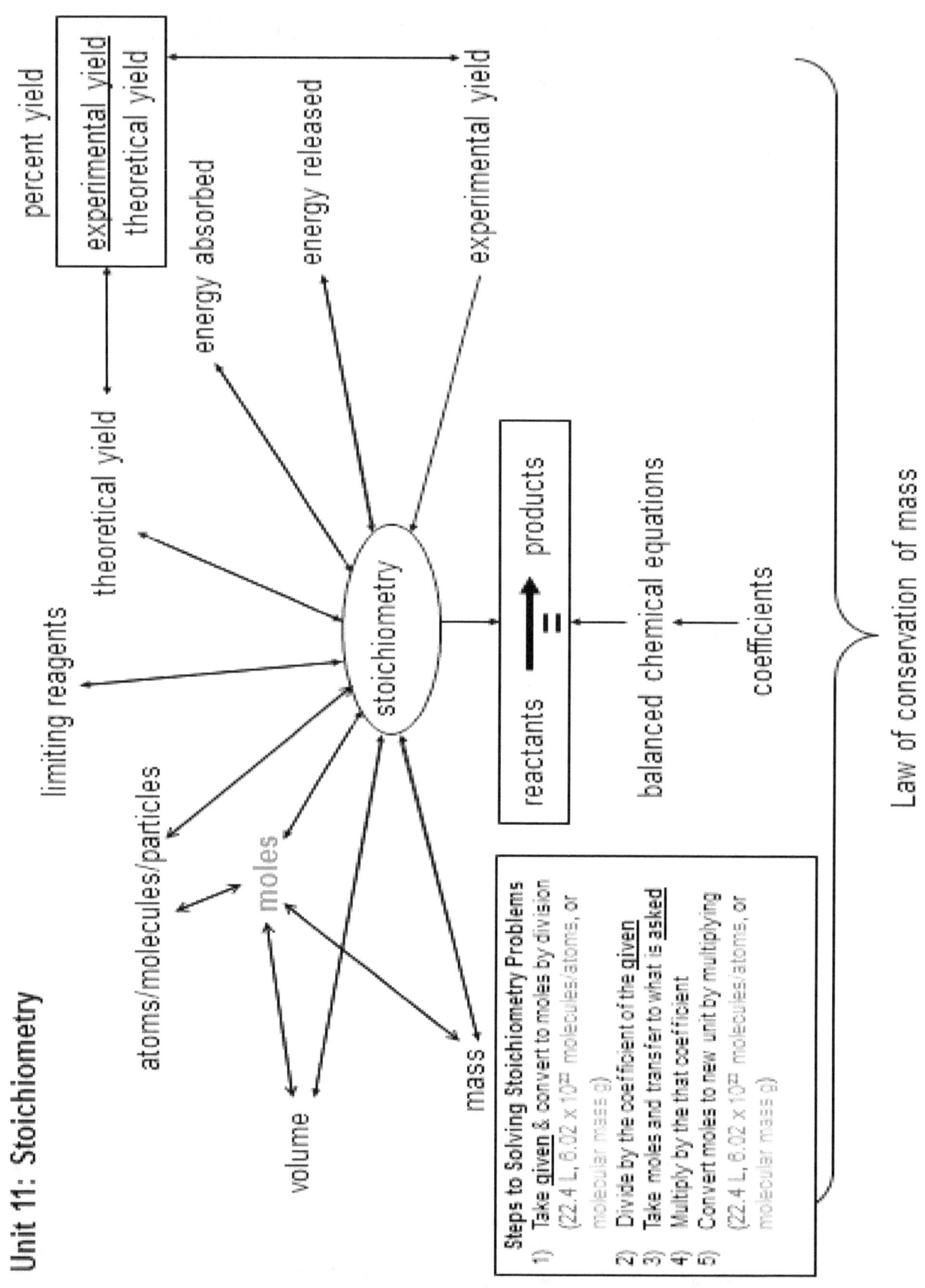

Conservation of Mass in Equations

Directions:

Draw lines from the elements in the reactants (the left side of the arrow) to the elements in the products (the right side of the arrow) for each balanced chemical equation to show how elements do not go away in a chemical reaction; they just get rearranged. An example is done below.

Example: $2H_2 + CO \rightarrow CH_3OH$

1) $2H_2O \rightarrow 2H_2 + O_2$

2) $HCl + NaHCO_3 \rightarrow CO_2 + H_2O + NaCl$

3) $C_6H_{12}O_6 + 6O_2 \rightarrow 6CO_2 + 6H_2O$

4) $CH_4 + 2O_2 \rightarrow CO_2 + 2H_2O$

5) $2H_2 + 2O_2 \rightarrow 2H_2O$

6) $C_2H_5OH + 3O_2 \rightarrow 2CO_2 + 3H_2O$

7) $6CO_2 + 6H_2O \rightarrow C_6H_{12}O_6 + 6O_2$

8) $2LiOH + CO_2 \rightarrow Li_2CO_3 + H_2O$

9) $(NH_4)_2Cr_2O_7 \rightarrow Cr_2O_3 + N_2 + 4H_2O$

10) $4NH_3 + 5O_2 \rightarrow 4NO + 6H_2O$

Questions:

1) How do these equations show mass is conserved?

2) How can a reaction have different reactants from its products but conserve mass?

3) What do the numbers in front of the formulas represent?

4) How can you tell how many elements there are in an equation?

5) Which careers use the information in the conservation of mass?

Conservation of Life: Photosynthesis and Respiration

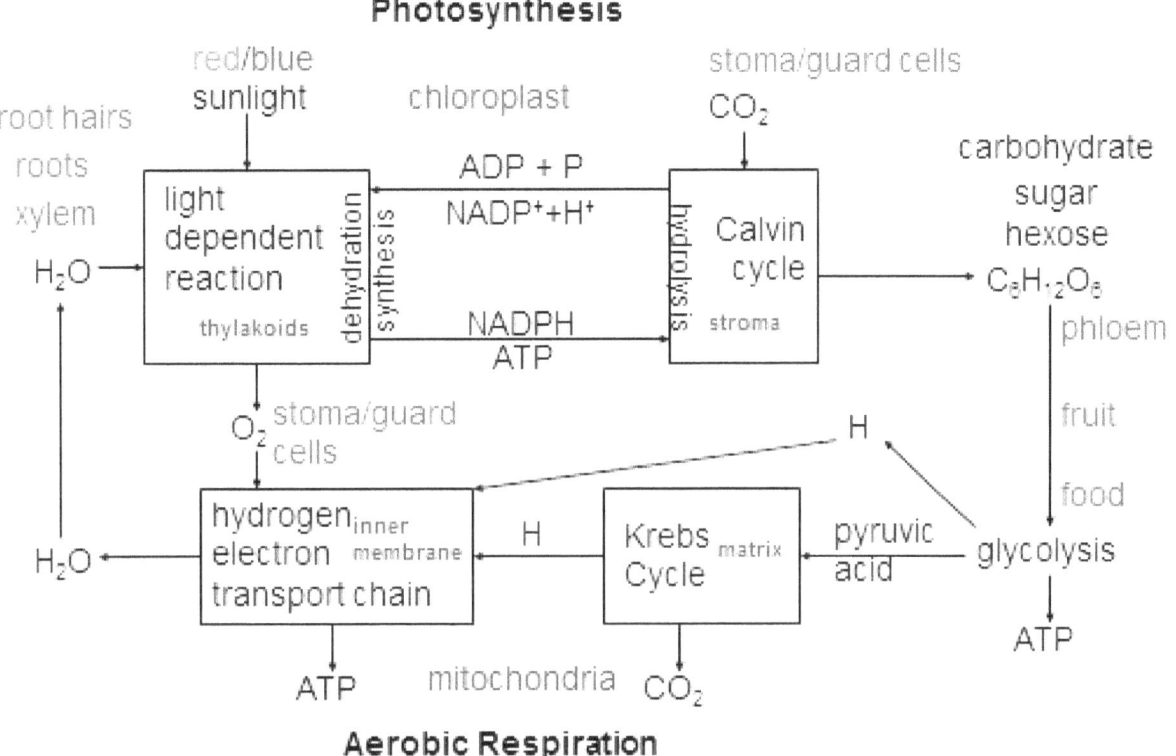

Directions:

Use the diagram above to answer the following questions.

1) Write the equation for photosynthesis and balance the reaction.

2) Write the equation for aerobic respiration and balance the reaction.

3) In both equations, trace where each element of the reactants go to make the products.

4) How are the two reactions similar to each other?

5) How does the conservation of mass, in this case, show that life has balance?

6) Plants and algae go through both photosynthesis and respiration. Animals only go through respiration. What would happen to life on Earth if we lost the plants and algae?

7) Keeping this in mind, which do you think formed first: the process of aerobic respiration or photosynthesis? Explain why.

Conservation of Mass

Directions and Questions:

You will need a **scale**, a **Ziploc bag**, **baking soda**, **vinegar**, and a **disposable pipette**. Looking at the materials and lab we will be using, what are the safety precautions we should take to protect ourselves and materials during the investigation?

1) Take a small Ziploc bag and put a spoonful of baking powder in it. Suck up some vinegar into the pipette and place the pipette in the bag. Seal the bag shut.
2) Find the mass of the bag and its contents before the reaction. What is the mass?

3) Squeeze the vinegar from the pipette into the baking soda. What do you observe?

4) Find the mass of the bag and its contents after the reaction. What is the mass?

5) Why do you think we made the reaction take place inside a sealed bag?

6) What did you see and feel that let you know a chemical reaction took place?

7) Was it an endothermic or exothermic reaction? How did you know?

8) Compare the mass of the bag and its contents before the reaction and after.

9) Why did the results come out as they did?

10) Use your textbook or the internet to find how the conservation of mass is stated, and write it here.

11) How did this reaction show the conservation of mass?

12) What is the chemical formula for vinegar or Acetic Acid?

13) What is the formula for baking soda or sodium bicarbonate?

The Law of Conservation of Mass

Directions and Questions:

1) You will need **safety goggles**, an **apron**, a **graduated cylinder**, 10 mL of **potassium iodide solution** in a **250 mL Erlenmeyer flask**, a **test tube** half full with a **lead nitrate solution** placed in the Erlenmeyer flask so that it does not mix with the potassium iodide, and a **rubber stopper** sealing the flask. **Looking at the materials and lab we will be using, what are the safety precautions we should take to protect ourselves and materials during the investigation?**

2) Measure the mass of the Erlenmeyer flask and its contents. What is it?

3) Carefully and slowly invert the flask, letting the test tube's contents mix with the flask's contents while keeping the stopper sealed tight. What did you observe?

4) What did you see that indicated a chemical reaction took place?

5) Measure the mass of the flask and its contents. What is it?

6) How did this reaction show the conservation of mass?

7) Why did we seal the flask?

The "Dozen" Lab

Directions and Questions:

You will need 12 **M&Ms**, 12 **Tic Tacs**, 12 **Skittles**, 12 **Peanut M&Ms**, and a **scale**. Looking at the materials and lab we will be using, what are the safety precautions we should take to protect ourselves and materials during the investigation?

1) Find the mass of 12 M&Ms, 12 Tic Tacs, 12 Skittles, and 12 Peanut M&Ms. Record the masses you found on the board in the front of the room with the other groups in your class.
2) Find a class average for the masses of the four candies. Record them in Data Table 1 below.

Data Table 1

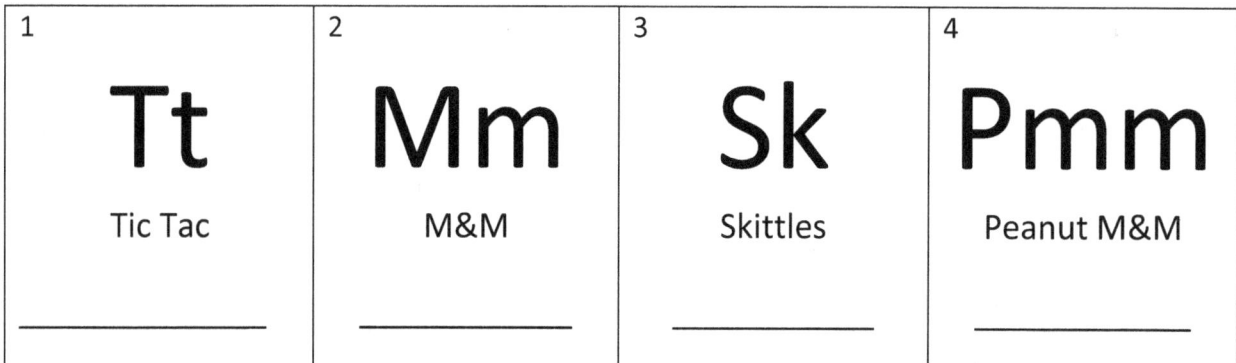

The number you wrote above is the **dozen mass** of each candy.

You can use the dozen mass to determine how many dozen candies are in a given sample. Use the equation: ***Mass of Sample ÷ dozen mass = number of dozen***

3) Using the dozen mass values from Data Table 1 to calculate the number of dozens in:
    a. 1000 g of Tic Tac

    b. 500 g of M&M

c. 250 g of Skittles

d. 3000 g of Peanut M&M

4) Now find out how many pieces of candy are in each sample by using the equation:
   Mass of Sample ÷ Dozen Mass x 12 = number of candies
   a) 450 g Tic Tac

   b) 2500 g M&M

   c) 3350 g Skittles

   d) 743 g Peanut M&M

In Chemistry, we deal with things much smaller than pieces of candy; we count atoms. Therefore our measurement is a much bigger number we call a mole. One mole equals 6.02×10^{23} atoms of something, but it works just like the dozen.

$$1 \text{ mole} = 6.02 \times 10^{23}$$

Instead of the Data Table of candy, we will use the Periodic Table of Elements, which has the average molar mass at the bottom of each square for each element.

5) Use the equation to solve how many moles are in a sample of elements:
   Mass of sample ÷ molar mass = number of moles
   a) 250 g of Magnesium (Mg)

   b) 35 g of Nitrogen (N)

c) 500 g of Lead (Pb)

d) 62 g of Carbon (C)

6) Since we know that one mole contains 6.02×10^{23} atoms, once we know how many moles there are, we can multiply by this number to find out how many atoms there are of that element. We can use the equation below to find the number of atoms for that sample:

Mass of Sample ÷ molar mass x 6.02×10^{23} = number of atoms

a) 500 g of Sodium (Na)

b) 430 g of Sulfur (S)

c) 75 g of Oxygen (O)

d) 84 g of Gold (Au)

Limiting Reactants

Directions and Questions:

Use the **internet** and your **textbook** to find <u>limiting reactants</u> and <u>excessive reactants</u>. Use what you find to answer the questions that follow.

1) You needed the following list to make a batch of cookies.
    a. 2.5 cups of flour
    b. 1 teaspoon of salt
    c. 2 eggs
    d. 2 sticks of butter
    e. 1 teaspoon of vanilla
    f. 1 cup of sugar
    g. 1 teaspoon of baking soda

    You have the following in your kitchen.

    a. 7 cups of flour
    b. 3 cups of salt
    c. 12 eggs
    d. 4 sticks of butter
    e. 12 teaspoons of vanilla
    f. 6 cups of sugar
    g. 2 cups of baking soda

    How many batches of cookies can you make?

    a. What is the limiting reactant?

    b. What are the excessive reactants?

7) In photosynthesis, a plant has 44 g of CO_2 and 32 g of H_2O available; what is the limiting reactant? **$6CO_2 + 6H_2O \rightarrow C_6H_{12}O_6 + 6O_2$**

a. What is the excessive reactant?

8) In the equation $3H_2 + N_2 \rightarrow 2NH_3$, if you have 4.83 moles of hydrogen and 8.49 moles of nitrogen, how much ammonia can be formed?

   a. What is the limiting reactant?

   b. What is the excessive reactant?

9) When 6 g of Sulfuric Acid reacts with 5 g of aluminum in the balanced equation **$3H_2SO_4 + 2Al(OH)_3$ to make $6H_2O + Al_2(SO_4)_3$**; how much water is formed?

   a. Which is the limiting reactant?

   b. Which is the excessive reactant?

Virtual Investigations that go with Stoichiometry

ExploreLearning.com:

    Stoichiometry Gizmo

    Limiting Reactants Gizmo

    Moles Gizmo

    Balancing Chemical Equations

Phet.colorado.edu:

    Reactants Products and Leftovers

    Balancing Chemical Equations

Physicsclassroom.com/Concept-Builders/Chemistry:

    Stoichiometry Relationships

    Mole Conversions

Unit 12: Acids & Bases

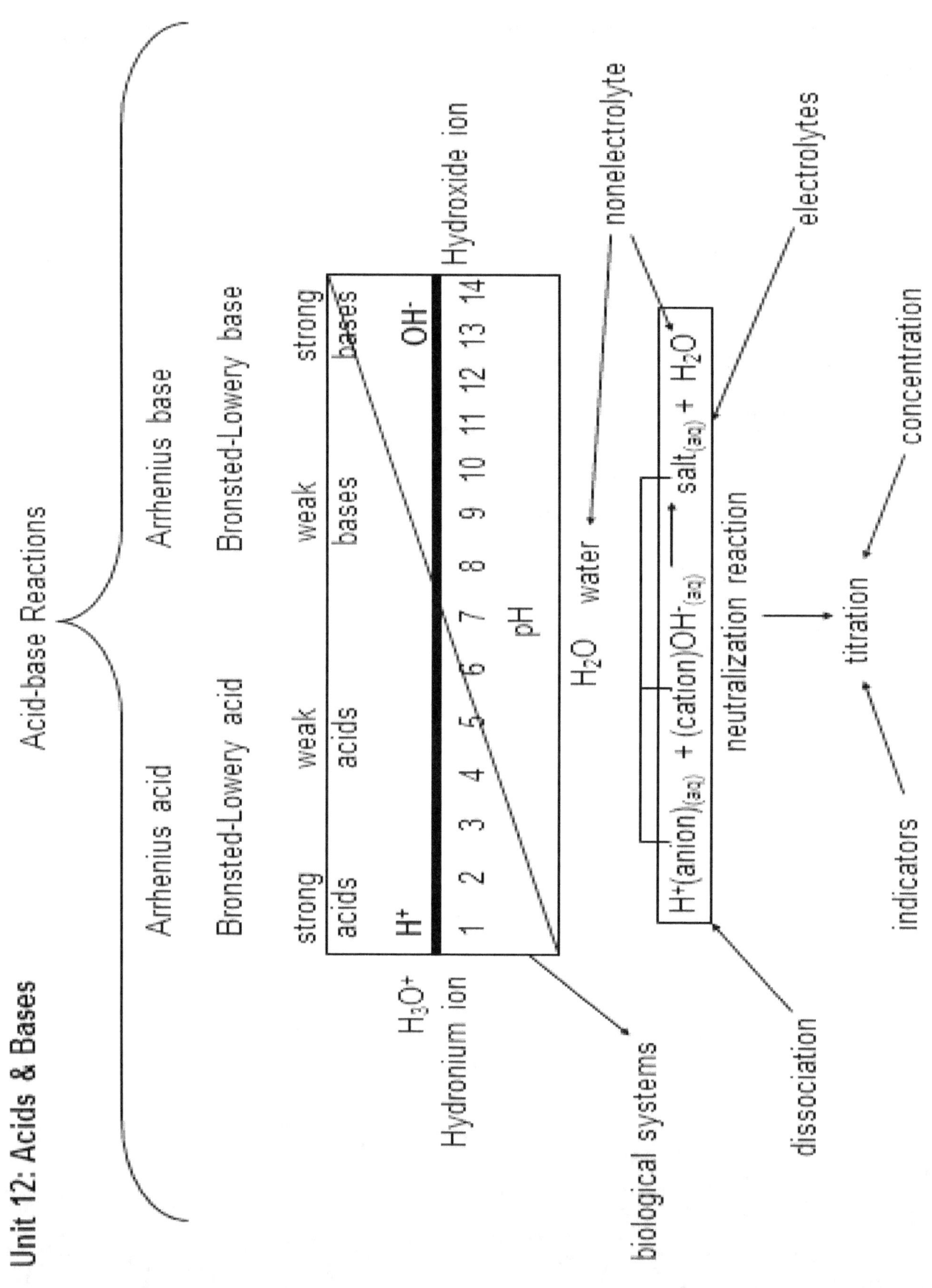

Which is an Acid and Which is a Base?

Directions:

Acids have H+ ions, and bases have OH- ions. Look at the formulas of some common acids and bases and see if you can tell them apart. Notice positive ions are written on the left side of the formula, and negative ions are written on the right side of the formula. Next to each formula, write down whether you think it is an acid or a base.

1) $Ca(OH)_2$

2) H_2SO_4

3) HCl

4) $NaOH$

5) $HC_2H_3O_2$

6) NH_4OH

7) KOH

8) H_2CO_3

9) $Ba(OH)_2$

10) $CsOH$

11) HNO_3

12) HBr

13) $Mg(OH)_2$

14) HI

15) $LiOH$

16) $Al(OH)_3$

17) $RbOH$

18) $HClO_4$

19) $HClO_3$

20) $Fe(OH)_2$

21) HBO_3

22) H_2S

23) $Zn(OH)_2$

24) HNO_2

25) $Fe(OH)_3$

26) HPO_3

27) $HC_2H_3O_2$

28) H_2SiO_3

Characteristics of Acids and Bases

Directions:

You will need **bottled water**, a **sink**, and put **vinegar, club soda, lemon juice, baking soda, soft soap,** and **laundry detergent** in **tiny cups** or **beakers**. Looking at the materials and lab we will be using, what are the safety precautions we should take to protect ourselves and materials during the investigation?

1) **Acids** are **sour,** and **bases** are **bitter** to the taste. **Bases feel slippery,** and **acids do not**. For each cup, separately dip your finger in and rub it on your thumb to find the feel (slippery or not slippery), and then lightly lick your finger to get a taste (bitter or sour). You can use the bottled water to wash your mouth out to get rid of the taste and use the sink to wash your fingers.
2) Circle the results in Data Table 1 below.

Data Table 1

Substance	Feel	Taste
Vinegar	Slippery or Not Slippery	Bitter or Sour
Club Soda	Slippery or Not Slippery	Bitter or Sour
Lemon Juice	Slippery or Not Slippery	Bitter or Sour
Baking Soda	Slippery or Not Slippery	Bitter or Sour
Soft Soap	Slippery or Not Slippery	Bitter or Sour
Laundry Detergent	Slippery or Not Slippery	Bitter or Sour

Questions:

1) Are cleaning agents acids or bases? Explain.

2) Why do lemon juice and some sodas taste sour?

Which will Corrode a Nail?

Directions:

You will need **two nails**, a small **bottle of coke** (you could also try **orange juice**), and a small bottle of **clear liquid soap. Looking at the materials and lab we will be using, what are the safety precautions we should take to protect ourselves and materials during the investigation?**

1) Acids tend to corrode metal, and bases do not. Coke has multiple acids in it, and soap is a base. **Hypothesis:** Which do you think will corrode the nail?

2) Open the lids of both the coke and the soap, place a nail in each one, and tightly fix the lids back on each.
3) Check them each day for two weeks to look for any signs of corrosion.
4) Fill in Data Table 1 below.

Data Table 1

Days	Corrosion of Nail in Coke	Corrosion of Nail in Soap
1	Yes or No	Yes or No
2	Yes or No	Yes or No
3	Yes or No	Yes or No
4	Yes or No	Yes or No
5	Yes or No	Yes or No
6	Yes or No	Yes or No
7	Yes or No	Yes or No
8	Yes or No	Yes or No
9	Yes or No	Yes or No
10	Yes or No	Yes or No
11	Yes or No	Yes or No
12	Yes or No	Yes or No
13	Yes or No	Yes or No
14	Yes or No	Yes or No

Question:

1) Did either show signs of corrosion in two weeks? If so, how?

2) Why do you think the pH of tap water is kept just over 7?

3) Why do you think soaps are good for cleaning metals?

4) Lemons contain lots of citric acid. Do you think lemon juice would corrode a nail? Explain.

5) Write a balanced equation for the reaction that would take place between citric acid and iron III.

A Homemade Indicator

Directions:

You will need **safety goggles**, an **apron**, a **500 mL beaker**, five **small beakers**, **water**, **red cabbage**, a **hotplate**, **shampoo**, **grapefruit juice**, **Sprite**, **milk**, and **ammonia**. Looking at the materials and lab we will be using, what are the safety precautions we should take to protect ourselves and materials during the investigation?

1) Place about 300 mL of water in the 500 mL beaker with some leaves of purple cabbage. Boil it until the water turns purple.
2) In 5 small beakers, separately place shampoo, grapefruit juice, Sprite, milk, and ammonia.
3) The red cabbage juice is red in most acids and blue-purple in most bases. Use a pipet to mix the red cabbage juice with each substance to determine whether it is **acid** or a **base**.

Results:

4) What is shampoo?

5) What is grapefruit juice?

6) What is Sprite?

7) What is milk?

8) What is ammonia?

Observing Acid Relief

Directions and Questions:

You will need **100% purple grape juice**, a **beaker**, **water**, and an **antacid tablet. Looking at the materials and lab we will be using, what are the safety precautions we should take to protect ourselves and materials during this investigation?**

1) Take a beaker and fill it halfway up with a solution of water and purple grape juice. The pigment in grape juice is a natural indicator like red cabbage juice. It turns red in acids and a grayish blue in a base. What color is the grape juice in the water solution?

2) Is the solution an acid or base?

3) When stomach acid gets too strong, we take antacid tablets to calm our stomach down. Drop an antacid tablet into the water and grape juice solution. What do you notice happening?

4) Is the ending solution an acid or base?

5) How does an antacid tablet help calm our stomach or acid reflux?

Simple Biology Investigations   Seven Sides Publishing

Acid or Base Grape Juice Indicator

Directions:

You will need **safety goggles**, an **apron**, a **pipette**, **100% purple grape juice**, seven tiny **beakers**, **vinegar**, **ammonia**, **lemon juice**, **Sprite**, **drain cleaner**, **detergent**, a **baking soda solution**, **litmus blue paper**, **litmus red paper**, **Universal indicator pH paper**, and a **pH meter** attached to an **interface** connected to a **computer** with **Logger Pro**. Looking at the materials and lab we will be using, what are the safety precautions we should take to protect ourselves and materials during the investigation?

1) Fill each of the seven beakers with one of these substances: vinegar, ammonia, lemon juice, Sprite, drain cleaner, detergent, and baking soda.
2) Use the red and blue litmus papers, dip them into each beaker, and write down what color they turn in Data Table 1.
3) Take the universal indicator pH paper, dip them into each beaker, and write down what pH the color indicated in Data Table 1.
4) Gently stir the pH meter in each beaker to see what the Logger Pro indicates is the pH of each solution. Be patient; it takes some time for the pH meter to stabilize. Make sure to rinse the pH meter between each beaker measurement. Write the pH measurement in Data Table 1.
5) Now take a pipette full of purple grape juice, squirt it into each beaker, and write down the color it makes in each of the solutions in Data Table 1.

Data Table 1

Solution	Blue Litmus	Red Litmus	Universal Indicator Paper	pH meter	Purple Grape Juice	Acid or Base?
Vinegar						
Ammonia						
Lemon juice						
Sprite						
Drain cleaner						
Detergent						
Baking soda						

Questions:

1) A pH below 7 indicates an acid; a pH above 7 indicates a base. What color does litmus paper turn when a solution is an acid?

2) What color does litmus paper turn when a solution is a base?

3) Determine if each substance is an **acid** or a **base** by filling in the last column in Data Table 1.

4) Which of the substances was the strongest acid (had the lowest pH)?

5) Which of the substances was the strongest base (had the highest pH)?

6) What color did the purple grape juice turn in the solution if it was an acid?

7) What color did the purple grape juice turn the solutions if it was a base?

8) Which of the indicators we used was most like the grape juice indicator? Explain.

9) Which of the indicators told you the most information? Explain.

10) Why do you think the purple grape juice is red when it is diluted?

Grape Juice Titration

Directions and Questions:

You will need **safety goggles**, an **apron**, a **50 mL burette**, **100% grape juice**, a **.1 M NaOH solution**, a **250 mL beaker**, a **graduated cylinder**, and a **stirring rod**. **Looking at the materials and lab we will be using, what are the safety precautions we should take to protect ourselves and materials during the investigation?**

1) We want to find the molarity of the citric acid ($C_6H_8O_7$) in grape juice. Measure 50 mL of grape juice and place it in the 250 mL beaker. The neutralization reaction is below:

    $C_3H_5O(COOH)_3 + 3NaOH \longrightarrow Na_3C_3H_5O(COO)_3 + 3H_2O$

2) Put 50 mL of .1M NaOH into the 50 mL burette. Stirring the grape juice slowly, add NaOH slowly from the burette until you see a temporary color change in the grape juice. Then add NaOH drop by drop until the grape juice stays that color other than red. How many mL of NaOH was used to neutralize the solution?

3) Calculate the number of moles of NaOH that were used.

4) How many moles of $C_6H_8O_7$ in grape juice were used?

5) Calculate the concentration of the acid.

Making a pH Scale

Directions:

You will need **safety goggles**, an **apron**, a **microwell plate** with at least 12 wells, four **pipettes**, a **1 M Hydrochloric Acid solution**, a **1 M Sodium hydroxide solution**, **distilled water**, a **universal indicator**, a **white sheet of paper**, **distilled water**, **lemon juice**, **milk**, and **liquid soap**. Looking at the materials and lab we will be using, what are the safety precautions we should take to protect ourselves and materials during the investigation?

1) Place the well plate over the white paper. Use one pipet to add nine drops of distilled water in wells 2-11.
2) Use another pipet to add ten drops of hydrochloric acid to well 1, then rinse with water. Transfer one drop of HCl to well 2. Whatever is left in the pipet, put back to well 1. Then mix well 2.
3) Then transfer one drop from well 2 to well 3. Whatever is left in the pipet, put back into well 2. Then mix well 3.
4) Repeat this process until you finish with well 6.
5) Use another pipet to add ten drops of Sodium Hydroxide to well 12, then rinse with water. Transfer one drop of NaOH to well 11. Whatever is left in the pipet, put back to well 12. Then mix well 11.
6) Then transfer one drop from well 11 to well 10. Whatever is left in the pipet, put back into well 11. Then mix well 10.
7) Repeat this process until you finish with well 8.
8) Add ten drops of distilled water into well 7
9) Using a different pipet add one drop of universal indicator into each well 1-12.
10) You now have an approximate pH value key for the universal indicator. See how closely it matches the color key that comes with the universal indicator. Well 1 is a pH of 1, well two is a pH of 2, and so on up to 12 because of how we diluted each well.
11) In unused wells with a rinsed pipet, place ten drops of lemon juice in one well, ten drops of milk in another well, and ten drops of liquid soap in another. Make sure you rinse the pipet each time before transferring a different substance so they do not contaminate the wells.
12) Place a drop of universal indicator in each well for the lemon juice, milk, and soap.
13) Notice the color they change and use wells 1-12 to estimate the pH of each.

Questions:

1) What pH values are acids?

2) What pH values are bases?

3) Distilled water is neutral. What pH value is distilled water?

4) What is the pH for Lemon juice?

5) Is lemon juice an acid or a base?

6) What is the pH of milk?

7) Is milk an acid or a base?

8) What is the pH for soap?

9) Is soap an acid or a base?

10) Acids tend to corrode metals. Why would soap have this pH?

11) What color/pH do you think tap water will be? Explain why?

   a. Test it; what color/pH was it?

12) Explain why the well plates we made acted like a pH meter?

13) What is a universal indicator?

Virtual Investigations that go with Acids and Bases

ExploreLearning.com:

pH Analysis Gizmo

pH Analysis: Quad Color Indicator Gizmo

Titration Gizmo

Phet.colorado.edu:

Acid-Base Solutions

pH Scale

pH Scale Basics

physicsclassroom.com/Concept-Builders/Chemistry:

Which One Doesn't Belong? Acid-Base Properties

Bronsted-Lowry Model of Acids and Bases

pH and pOH

Chemistry and IPC TEKS & NGSS Correlations

Nature of Science Concept Map Chem c 1ABH 2D 3AB 4AB

Focus on the Process Chem c 1ABCDEFG 2AC 3ABC 4A

Measurement Lab Chem c 1ABCDEF 2AB 3AB 4A

Patterns in Pennies Chem c 1ABCDEF 2BCD 3ABC 4A

Virtual Introductory Investigations Chem c 1ABEFG 2ABCD 3ABC 4AB

Properties of and Changes in Matter Concept Map Chem c 5A, IPC c 7A 8AB; HS-PS1-1

Measure and Calculate Density Chem c 1AC 2EFGHI 3A 4AB, IPC c 1AC 2BCDE 3A 6ABC; HS-PS1-1

The Density of Oddly Shaped Objects Chem c 1ABCDEF 2BC 3AB 4A 5A, IPC c 1ABCDEF 2BC 3AB 4A 7B; HS-PS1-1

Which is Denser? Chem c 1AC 2EHI 3A 4AB, IPC c 1AC 2BDE 3A 6ABC; HS-PS1-1

Extensive and Intensive Properties Chem c 1ABCDEF 2BC 3AB 4A 5A, IPC c 1ABCDEF 2BC 3AB 4A 7B

Physical and Chemical Changes Chem c 1ABCDEF 2B 3AB 4A 5A, IPC c 1ABCDEF 2B 3AB 4A 6D 7BC 8A

Virtual Properties and Changes in Matter Investigations Chem c 1ABEFG 2ABCD 3AB 4AB 5A 8AB, IPC c 1ABEFG 2ABCD 3AB 4AB 7B 8A; HS-PS1-1

Gas Laws Concept Map Chem c 10AB; HS-PS3-2

Observing Molecular Motion Chem c 1ABCDEF 3AB 4A 10A , IPC c 1ABCDEF 3AB 4A 6D; HS-PS3-2

Observing Boyles Law Chem c 1ABCDGH 2AB 3AB 4A 10B; HS-PS3-2

Relationship Between Temperature, Volume, and Pressure: Charles Law and Gay-Lussac's Law Chem c 1ABCDGH 2AB 3AB 4A 10B; HS-PS3-2

Ideal Gas Law Calculations Chem c 1ABH 2BC 3AB 4A 10BC; HS-PS3-2

Virtual Gas Laws Investigations Chem c 1ABEFGH 2ABCD 3ABC 4A 10AB; HS-PS3-2

Energy and Phase Changes Concept Map Chem c 10A 13ABD; HS-PS3-2

The Four Laws of Thermodynamics Chem c 1ABE 3AB 4A 13A, IPC c 1ABE 3AB 4A 6D; HS-PS3-2

Student Atomic Motion Chem c 1ABCG 2AD 3AB 4A 10A, IPC c 1ABCG 2AD 3AB 4A 6D; HS-PS3-2

Seeing the Heating Curve Chem c 1ABCDEF 2B 3AB 4A 10A, IPC c 1ABCDEF 2B 3AB 4A 6D; HS-PS3-2

Calorimetry Lab Chem c 1ABCDEF 2BCD 3AB 4A 13BD, IPC c 1ABCDEF 2BCD 3AB 4A 7C; HS-PS3-2

Observing and Calculating Change in Energy Chem c 1ABCDE 2BC 3AB 4A 13BD; HS-PS3-2

Convection in Liquids and Gases Chem c 1ABCDFG 3AB 4A 10A 13A, IPC c 1ABCDFG 3AB 4A 6D; HS-PS3-2

Observing Conduction Convection and Radiation Chem c 1ABCDF 3AB 4A 10A 13A, IPC c 1ABCDF 3AB 4A 6D 7C; HS-PS3-2

Energy Transformation Balls Chem c 1ABCDEH 3AB 4A 10A 13A, IPC c 1ABCDEH 3AB 4A 6D; HS-PS3-2

Virtual Energy, Phase Changes, and Calorimetry Investigations Chem c 1ABEFG 2ABCD 3AB 4A 10A 13ABD, IPC c 1ABEFG 2ABCD 3AB 4A 6D 7C; HS-PS3-2

Pure Substance and Mixtures Concept Map Chem c 10AB 11ABCDE; HS-PS1-3

Speed of Dissolving Solutes Lab Chem c 1ABCD 3AB 4A 10AB 11C, IPC c 1ABCD 3AB 4A 7F; HS-PS1-3

Heat and Saturating Solutions Chem c 1ABCD 3AB 4A 10AB 11BCD, IPC c 1ABCD 3AB 4A 7F; HS-PS1-3

Sugar or Salt Chem c 1ABCDEFH 2BC 3AB 4A 11ABEF, IPC c 1ABCDEFH 2BC 3AB 4A 7F; HS-PS1-3

Boiling Points of Solutions Chem c 1ABCDEFH 2B 3AB 4A 11C 13A, IPC c 1ABCDEFH 2B 3AB 4A 6D 7C; HS-PS1-3

Making Ice Cream Chem c 1ABCDEF 2BCD 3AB 4A 11C 13A, IPC c 1ABCDEF 2BCD 3AB 4A 6D 7C; HS-PS1-3

The Solubility of Gas in a Liquid Chem c 1ABCD 3AB 4A 10AB 11C, IPC c 1ABCD 3AB 4A 7F; HS-PS1-3

Using Beer's Law Chem c 1ABCDEFG 2BC 3AB 4A 11EF; HS-PS1-3

Elements Compounds and Mixture Research Chem c 1AB 3AB 4A 11D, IPC c 1AB 3AB 4A 7B

Elements Compounds and Mixtures Chem c 1ABCDF 2B 3AB 4A 11D, IPC c 1ABCDF 2B 3AB 4A 7BF

Separating Mixtures Chem c 1ABCD 2D 3AB 4A, IPC c 1ABCD 2D 3AB 4A 7BF

Separating Pigments Chem c 1ABCDG 2AB 3AB 4A, IPC c 1ABCDG 2AB 3AB 4A 7BF

50 + 50 does not equal 100 Chem c 1ABCDEG 2ABC 3AB 4A, IPC c 1ABCDEG 2ABC 3AB 4A 7B

Percent Sugar in Bubble Gum Chem c 1ABCDEF 2BC 3AB 4A 8C, IPC c 1ABCDEF 2BC 3AB 4A 7BF

Virtual Pure Substance and Mixtures Investigations Chem c 1ABEFG 2ABCD 3AB 4A 10AB 11ABCDEF, IPC c 1ABEFG 2ABCD 3AB 4A 6D 7BF; HS-PS1-3

Atomic Structure and Nuclear Concept Map Chem c 6ABCD 14AB; HS-PS1-1

Simple History of the Atom Chem c 1ABEG 3AB 4A 6AB, IPC c 1ABEG 3AB 4A 7ABDE; HS-PS1-1

Picture of Atoms Chem c 1ABGH 2A 3ABC 4ABC 6AB, IPC c 1ABCH 2A 3ABC 4ABC 7A; HS-PS1-1

Scale Model of a Hydrogen Atom Chem c 1ABCD 2AB 3ABC 4A 6AB, IPC c 1ABCD 2AB 3ABC 4A 7A; HS-PS1-1

Model of an Atom Showing the Illusion Chem c 1ABCDG 2AB 3ABC 4A 6AB, IPC c 1ABCDG 2AB 3ABC 4A 7A; HS-PS1-1

Building Bohr Models Chem c 1ABCDEFG 2ABCD 3AB 4A 5ABC 6BDE, IPC c 1ABCDEFG 2ABCD 3AB 4A 7AB; HS-PS1-18

Half-life of Pennies Chem c 1ABCDEFG 2ABCD 3AB 4A 14A, IPC c 1ABCDEFG 2ABCD 3AB 4A 8C; HS-PS1-3

Calculating Nuclear Half-life Decay Chem c 1ABEFG 2BC 3AB 4A 14A, IPC c 1ABEFG 2BC 3AB 4A 8C; HS-PS1-3

Nuclear Decay Chem c 1ABEG 3AB 4A 14A, IPC C 1ABEG 3AB 4A 7A; HS-PS1-3

Nuclear Isotopes Chem c 1AB 3AB 4AC 14AC, IPC c 1AB 3AB 4AC 5D 8C; HS-PS1-13

Nuclear Fission and Fusion Chem c 1AB 3AB 4AC 14BC, IPC c 1AB 3AB 4AC 5D 8CD; HS-PS1-3

Nuclear Chain Reactions Chem c 1ABCDEFG 2AB 3AB 4A 14C, IPC c 1ABCDEFG 2AB 3AB 4A 8C; HS-PS1-3

Nuclear Reactor Chem c 1AB 3AB 4AC 14C, IPC c 1AB 3AB 4AC 5D 8C; HS-PS1-3

Smoke Detector Chem c 1AB 3AB 4AC 14C, IPC c 1AB 3AB 4AC 8C; HS-PS1-3

Applications of Nuclear Phenomena Chem c 1ABEG 3AB 4A 14C, IPC c 1ABEG 3AB 4A C; HS-PS1-3

Virtual Atomic Structure and Nuclear Investigations Chem c 1ABEFG 2ABCD 3AB 4AB 6ABCDE 14ABC, IPC c 1ABEFG 2ABCD 3AB 4AB 7AB 8C; HS-PS1-13

Electron Concept Map Chem c 6ABCDE; HS-PS1-1, 4-4

Niels Bohr's Contributions of Electrons Chem c 1ABEG 3AB 4A 5AB 6AB, IPC c 1ABEG 3AB 4A 7ABDE; HS-PS1-1, 4-34

Seats and Electron Configuration One Class Chem c 1ABEG 2A 3AB 5ABC; HS-PS1-1

Seats and Electron Configuration Many Classes Chem c 1ABEG 2A 3AB 5ABC; HS-PS1-1

Atomic Structure and Representations Chem c 1BFG 2BC 3AB 4A 5ABC 6D; HS-PS1-1

Summary of Electrons in each Energy Level Chem c 1AB 3AB 4AC 5ABC 6D; HS-PS1-1

Electron Basics Chem c 1AB 3AB 4ABC 5A 5AB, IPC c 1AB 3AB 4ABC 7DE; HS-PS1-1, 4-34

Uses of the Electromagnetic Spectrum Chem c 1ABEG 2B 3AB 4A 6C, IPC c 1ABEG 3AB 4A 5D 6F; HS-PS4-45

How we use Microwaves Chem c 1ABCDE 2B 3AB 4A 14C, IPC c 1ABCDEF 3AB 4A 5AD 6CDE 7C; HS-PS4-5

Virtual Electron Investigations Chem c 1ABEFG 2ABCD 3AB 4AB 6AC, IPC c 1ABEFG 2ABCD 3AB 4AB 7DE; HS-PS1-1. 4-345

Periodic Table Concept Map Chem c 5ABC 6D 7A, IPC c 7ABD; HS-PS1-1

Flame Test Chem c 1ABCDEF 2AB 3AB 4A 6C, 1PC c 1ABCDEF 2AB 3AB 4A 7D; HS-PS1-1

Metal or Nonmetal Chem c 1ABCDEF 2B 3AB 4A 5AB 7AD, IPC c 1ABCDEF 2B 3AB 4A 7AB; HS-PS1-16

Finding the Period in the Periodic Table Chem c 1ABEF 2AB 3AB 4A 5AB, IPC c 1ABEF 2AB 3AB 4A 7ABD; HS-PS1-1

Periodic Table Activity Chem c 1ABEFG 2AB 3AB 4A 5ABC 7A, IPC c 1ABEFG 2AB 3AB 4A 7ABD; HS-PS1-1

Making Lewis Dot Structures Chem c 1ABEG 2B 3AB 6E, IPC c 1ABEG 2B 3AB 7AB; HS-PS1-1

Making a Graphite Light Bulb Chem c 1ABCDEG 2AD 3AB 4A 5B 6ABC, IPC c 1ABCDEG 2AD 3AB 4A 6ADE 7ABCD; HS-PS1-1, 3-2

Virtual Periodic Table Investigations Chem c 1ABEFG 2ABC 3AB 4A 5ABC 6D 7AB, IPC c 1ABEFG 2ABC 3AB 4A 7AB; HS-PS1-1

Bonding Concept Maps Chem c 5B 6E 7ABCD; HS-PS1-12

Ionic Bonding Models Chem c 1ABCDG 2AB 3AB 4A 5B 6E 7ABCD 8C, IPC c 1ABCDG 2AB 3AB 4A 7AB; HS-PS1-12

Naming Ionic Compounds Chem c 1ABEF 2BC 3B 7B 12A, IPC c 1ABEF 2BC 3B 7A; HS-PS1-12

Metallic Bond Research Chem c 1AB 3AB 4AC 7D, IPC c 1AB 3AB 4AC 7BC; HS-PS1-12, 3-2

Valence Shell Electron Pair Repulsion Theory Chem c 1AB 3AB 4AC 7C; HS-PS1-12, 3-5

Building a Model of a Water Molecule Chem c 1ABCDEG 3AB 4A 5B 7ACD, IPC c 1ABCDEG 3AB 4A 7ABC; HS-PS1-12, 3-5

How Does Rain Form? Chem c 1ABCDE 3AB 4A 7AD, IPC c 1ABCDE 3AB 4A 7ABC; HS-PS3-5

Molecular Geometry Chem c 1ABCDG 2AB 3AB 4A 5B 6E 7ABCD 8C, IPC c 1ABCDG 2AB 3AB 4A 7AB; HS-PS3-5

Naming Covalent Compounds Chem 1ABEF 2C 3B 7B, IPC c 1ABEF 2BC 3B 7A; HS-PS1-12

Empirical and Molecular Formulas Chem c 1AB 2BC 8CD; HS-PS1-12

Breaking Bonds Chem c 1ABCDE 2B 3AB 4A 7D, IPC c 1ABCDE 2B 3AB 4A 7AB; HS-PS1-123

The Conductivity of Electrolyte Mixtures Chem c 1ABCDEF 2A 3AB 4A 7D 11AB, IPC c 1ABCDEF 2A 3AB 4A 7AB; HS-PS1-12, 2-6, 3-2

Polarity of Water Chem c 1ABCDE 2B 3AB 4A 11A, IPC c 1ABCDE 2B 3AB 4A 7A; HS-PS1-1, 3-5

Checking Polarity Chem c 1ABCDE 2B 3AB 4A 11A, IPC c 1ABCDE 2B 3AB 4A 7A; HS-PS1-3, 3-5

Virtual Bonding Investigations Chem c 1ABEFG 2ABCD 3AB 4A 6E 7ABCD 11A, IPC c 1ABEFG 2ABCD 3AB 4A 7ABC; HS-PS1-123, 2-6, 3-25

Energy in Chemical Changes Concept Map Chem c 10A 13C; HS-PS1-4

Interpreting Activation Energy for a Reaction Chem c 1ABEG 3AB 4A 13C, IPC c 1ABEG 3AB 4A 6D; HS-PS1-4

Temperature and Reaction Rates Chem c 1ABCDE 2C 3AB 4A 10A 13C, IPC c 1ABCDE 2C 3AB 4A 7F; HS-PS1-45

Observing a Catalyst Chem c 1ABCDE 2B 3AB 4A 13C, IPC c 1ABCDE 2B 3AB 4A 7F; HS-PS1-45

Endothermic and Exothermic Reactions Chem c 1ABCDE 2B 3AB 4A 13C, IPC c 1ABCDE 2B 3AB 4A 7C 8A; HS-PS1-4

Endothermic and Exothermic Models Chem c 1ABEG 2B 3AB 4A 13C, IPC C 1ABEG 2B 3AB 4A 6D; HS-PS1-4

Virtual Energy in Chemical Changes Investigations Chem c 1ABEFG 2ABC 3AB 4A 10A 13C, IPC c 1ABEFG 2ABC 3AB 4A 7CF 8A; HS-PS1-45

Classification of Chemical Reactions Concept Map Chem c 7B 9AB; HS-PS1-27

Home Chemistry Chem c 1ABCDE 3AB 4A 7B 9AB, IPC c 1ABCDE 3AB 4A 8AB; HS-PS1-27

Types of Chemical Reactions Chem c 1ABCDE 3AB 4A 7B 9AB 12A, IPC c 1ABCDE 3AB 4A 8AB; HS-PS1-27

Removing Carbon from Sugar Chem c 1ABCDE 3AB 4A 7B 9AB 12A, IPC c 1ABCDE 3AB 4A 8AB; HS-PS1-27

Nonrenewable Resources Chart Chem c 1AB 3AB 4AC, IPC c 1AB 3AB 4AC 8CD; HS-PS1-2

Renewable Resources Chart Chem c 1AB 3AB 4AC, IPC c 1AB 3AB 4AC 8D; HS-PS1-2

Oxidation-Reduction Titration Chem c 1ABCDE 3AB 4A 7B 8ABCD 9B 11EF, IPC c 1ABCDE 3AB 4A 8AB; HS-PS1-27

Using Activity Series to Predict Reactions Chem c 3B 5BC 7A; HS-PS1-123

Virtual Classification of Chemical Reactions Investigations Chem c 1ABEFG 2ABC 3AB 4A 9ABCD, IPC c 1ABEFG 2ABC 3AB 4A 8AB; HS-PS1-1237

Stoichiometry Concept Map Chem c 9CD; HS-PS1-7

Conservation of Mass Equations Chem c 2C 9A, IPC 2C 8B; HS-PS1-7

Conservation of Life: Photosynthesis and Respiration Chem c 1ABG 2C 3AB 4A 9A, IPC 1ABG 2C 3AB 8B; HS-PS1-7

Conservation of Mass Chem c 1ABCDEG 2AB 3AB 4A 9C, IPC c 1ABCDEG 2AB 3AB 4A 8B; HS-PS1-7

The Law of Conservation of Mass Chem c 1ABCDEG 2AB 3AB 4A 9C, IPC c 1ABCDEG 2AB 3AB 4A 8B; HS-PS1-7

The "Dozen" Lab Chem c 1ABCDEFG 2ABC 3AB 4A 9C; HS-PS1-7

Limiting Reactants Chem c 1AB 2BC 3AB 4AC 9BCD; HS-PS1-67

Virtual Stoichiometry Investigations Chem c 1ABEFG 2ABC 3AB 4A 9CD, IPC c 1ABEFG 2ABC 3AB 4A 8B; HS-PS1-67

Acids and Bases Concept Map Chem c 9A 12ABCDE; HS-PS1-1257

Which is an Acid and Which is a Base? Chem c 3A 6B 12B, IPC c 3A 7B; HS-PS1-1

Characteristics of Acids and Bases Chem c 1ABCDEF 3AB 12B, IPC c 1ABCDEF 3AB 7C; HS-PS1-12

Which will Corrode a Nail? Chem c 1ABCDEFH 3AB 4A 9A, IPC 1ABCDEFH 3AB 4A 8AB; HS-PS1-2

A Homemade Indicator Chem c 1ABCDE 3AB 4A 12B, IPC 1ABCDE 3AB 4A 7AC 8A

Observing Acid Relief Chem c 1ABCDE 3AB 4A 12BC, IPC c 1ABCDE 3AB 4A 7ABC 8A

Acid or Base Grape Juice Indicator Chem c 1ABCDEF 2B 3AB 4A 12BC, IPC c 1ABCDEF 3AB 4A 7A 8A; HS-PS1-127

Grape Juice Titration Chem c 1ABCDE 2BC 3AB 4A 12ABD, IPC c 1ABCDE 2BC 3AB 4A 8A; HS-PS1-1267

Making a pH Scale Chem c 1ABCDEG 2ABC 3AB 4A 12ABCE, IPC c 1ABCDEG 2ABC 3AB 4A 7F 8A; HS-PS1-127

Virtual Acids and Bases Investigations Chem c 1ABEFG 2AB 3AB 4A 12ABCDE, IPC c 1ABEFG 2AB 3AB 4A 7AF 8A; HS-PS1-127

Equipment List for All of the Investigations

If you want to do all the labs in this manual, here is a list of all the equipment you will need in order of appearance.

Small Lego sets	Variety of beakers
Scales	Vinegar
Meter sticks	Baking soda
Temperature probes	Steel wool
Computers	Beaker tongs
Interfaces	Copper
Logger Pro software	Corroded copper
100 mL graduated cylinders	Hydrogen peroxide
Stopwatches	Liver
Rulers	Bananas
Pennies	Refrigerator and freezer
Penny rolls	Hotplates
Tap water	Food coloring
Distilled water	Large syringes and stopcocks
Safety goggles	Small marshmallows
Aluminum pans	Balloons
Paper	String
Candles	Erlenmeyer flasks and rubber stoppers
Long neck lighters	Rubber stopper assemblies
Matches	Gas pressure sensors &
Salt	Plastic tubing

Ring stands and clamps	Quart and gallon-sized Ziploc bags
Paper clips	Nesquik
Ammonium nitrate (s)	Milk
Urea (s)	Laser pointer
Sodium hydroxide (s)	Samples of granite countertop
Scoopulas	Cool Aid
Styrofoam cups	Pencils
Pepper	Chalk
Sugar cubes	Muddy water
Sandwich Ziploc bags	Magnets
Test tubes	Filter paper
Test tube holders/tongs	Wire strainers
Test tube racks	Marbles
Stirring rods	Sand
Glass bottles of soda	Iron filings
Measuring pipets	Granola
Pipets/eyedroppers	Pens
.4 M Nickle II Sulfate solution	Markers
Lens paper	Nail polish remover
Calorimeters	Rubbing alcohol
Aluminum foil	Chromatography paper
Gloves	Rubber stoppers with holes for test tubes
Crushed ice	Bubble gum
Spoons	Golf balls

- Film cases
- Red, white, and blue beads
- Periodic Table
- Dominos
- Plastic tubs
- Q-tips
- Bunsen burners
- Spark starters
- LiCl solution
- $CaCl_2$ solution
- KCl solution
- $CuCl_2$ solution
- $SrCl_2$ solution
- NaCl solution
- $BaCl_2$ solution
- Sulfur
- Charcoal
- Copper
- Glass jar and lid
- .2-.5 mm Graphite (mechanical pencil lead)
- 6-volt lantern batteries
- Molecular model kit
- Blue tac
- Wires with alligator clips
- Internet
- Batteries
- Battery packs
- Christmas lights
- Sink
- Paper towel
- Alka-Seltzer tablets
- Mentos
- Epsom salt
- Borax
- Laundry detergent
- Small cups
- Copper sulfate
- Nails
- Screws
- Ceramic bowls
- Lighter fluid
- Bleach
- 50 mL burette
- Oxidation-Reduction Potential sensors
- Magnetic stirrers
- Potassium iodide solution
- Lead nitrate

M&Ms

Tic Tacs

Skittles

Peanut M&Ms

Plastic bottles of Coke

Bottles of orange juice

Red cabbage

Shampoo

Grapefruit juice

Sprite

Ammonia

100% grape juice

Antacid tablets

Lemon juice

Drain cleaner

Red and blue litmus paper

Universal indicator pH paper

pH probe

1 M HCl solution

1 M NaOH solution

Aprons

Microwell plates

Universal indicator

Liquid soap

www.ingramcontent.com/pod-product-compliance
Lightning Source LLC
Chambersburg PA
CBHW080453220526
45465CB00006B/2262